Anomaly: Revolutionary Knowledge in Everyday Life

Jake Shannon

Anomaly: Revolutionary Knowledge in Everyday Life
Copyright © 2009 by Jake Shannon

Typeset by Joshua W

For more information, please visit www.JakeShannon.com

Printed in the United States of America. All rights reserved under International Copyright Law.

Cover and/or contents may not be reproduced in any manner without the express written consent of the author.

"Convictions are more dangerous enemies of truth than lies."
 –Friedrich Nietzsche

This book is dedicated to the joys of my life–Sondra and Leona and the pride, Dantzig. I look forward to a very long, healthy, and happy life together. I love you all!

ACKNOWLEDGEMENTS

"*If I have seen farther, it is only by standing on the shoulders of giants,*" said Sir Newton. The anomalies around me stir me to explore more and share what I see. As an essayist and rhetorician, I hunger to stir people up and challenge their beliefs. My attempt has been to bring together the truly profound insights of great thinkers, especially of these intellectual heroes of mine: William R. Corliss, Richard Dawkins, Daniel Ellsberg, Sam Harris, Seymour Hersch, F. A. Hayek, Aldous Huxley, Robert Jahn, Julian Jaynes, Thomas Kuhn, Sister Miriam Joseph, Ray McGovern, Friedrich Nietzsche, George Orwell, Thomas Paine, Karl Popper, Bertrand Russell, Thomas Szasz, Nassim Taleb, and Marcello Truzzi.

I thank my mother and father, and my wife Sondra for encouraging and supporting my development, especially as an anomalist. For providing a critical review, I am thankful to Amit Bhandari, Todd Ian Stark, and my editor Liza Joseph.

If I rouse in you an interest in the study of anomaly, I'll consider my endeavor a success. If you look around and recognize trope, I'll consider my attempt fruitful. The task of examining paradigms and integrating anomalies is enormous—not one man's task. My hope rests on those that are much smarter than me doing a better job in the future! I wish to thank you, the reader, for doing the work that may be required to digest the first third of this book and hope to reward you with the 'hair-raising' (my editor's description) remainder.

CONTENTS

Acknowledgements   ix

I   Prolegomena To Any Future Metaphor   1

1   Anomaly   3
    A Roadmap To This Book 7

2   But It's *Analogies* All The Way Down…   9
    The Mighty Pen 10
    The Conflict 12
    How Much Can We Know? 14
       Schemes and Tropes 19

3   First-Hand, In-Built, And Epistemic Hand-Me-Downs   21
    First-hand and Inbuilt Knowledge 22
    Epistemic Hand-Me-Downs 24
    Tacit and Explicit Knowledge 26
    Language and Thought 28
    A Note on the Mechanics of Communication 29

4   Reasoning: Our Tool For Learning   31
    Grammar: The Thing-As-It-Is-Symbolized 32
    Logic: The Thing-As-It-Is-Known 33
       A Note on Science, Falsification, and Skepticism . 35
       Errors, Cognitive Illusions, and Informal Fallacies 38
    Rhetoric: The Thing-As-It-Is-Communicated 41
       Polysemes 42

	Tropes or Figures of Speech	43
5	**Hypnosis: Rhetoric-In-Action**	**45**
	Myths	51
6	**Epistemic Hand-Me-Downs**	**55**
7	**Thought Control Vs. First Amendment**	**63**
	Law: The Rhetoric of Peaceful Liberty	66
8	**The Helplessness Paradigm**	**69**
	Psychiatric Trope	78
	The Institutionalization of Learned Helplessness	85
	The Pedagogy of Surrender	87
	There is No 'Conscience' Without 'Science'	88

II New Word Order – Paradigms Of Trope & Centralization 95

9	**Secret Paradigms**	**97**
10	**The Infidel Revolution**	**101**
	Religions Often Call For, and Get, Murder	104
	Religion is Not Always About God; It's About Power	105
	The Founding Fathers Were Not Christians	107
	The Nazis, However, Were Christian	113
	Nietzsche, Hitler's Adopted Philosopher, Was VERY Anti-Nazi	115
11	**Politics As Pro-Wrestling**	**121**
12	**The Centralization Of Symbology And The Paradigm Of Tyranny**	**135**
13	**Chance And Financial Anomalies**	**155**

14 The Anomalist	165
Recommended Resources	179
Bibliography	185
About Jake Shannon	205

Part I

Prolegomena To Any Future Metaphor

CHAPTER I

ANOMALY

Ever since men became capable of free speculation, their actions, in innumerable important respects, have depended upon their theories as to the world and human life, as to what is good and what is evil. (Russell 1945, 2)

Theories or models about the world and human life take various forms: religions, scientific theories, statistics, philosophical paradigms, or even the language we use every day. These models are representations of reality and are essentially a form of analogy, with some models delivering richer, more reliable information than others. Analogies are central to the way humans think. Language is steeped in tropes (that is, figures of speech)–some apparent, others not at all. For example, we use light-related analogies to talk about states of the intellect–brilliant, dull, clear, illuminating, to list only a few–and hardly notice it. Tropes and models have a deep influence on how we *see* reality. The *Book of Genesis*, the first book of the *Bible*, presents a model of how all that we see came into being; Charles Darwin's *The Origin of Species* presents another model of how all things living that we see today came into being–very different models that yield very different results, or rather consequences. As cognitive scientist Douglas Hofstadter says, analogy is "the lifeblood" of thought (D. Hofstadter 1982).

Analogies will always only be approximations; they're not reality. The menu is not the meal. With any theory, expectation, or model of reality, there exists a gap between what is real and what is representation–language, models, and paradigms. A trip to the carnival to watch a contortionist, acrobat, strong man, or stage hypnotist act presents unusual and amazing feats that are beyond normal everyday expectations. This gap between what is expected and what actually happens can challenge your worldview. This gap, the difference you observe between an analogy (models, paradigms, and trope) and the reality it attempts to describe, is anomaly.

The true scientist, philosophical truth-seeker, and noble rhetorician actively seek models and paradigms that better fit reality, but there are those that seek to manipulate models to their advantage, and you have probably experienced such anomalies–politicians who persuade you that they will implement effective governance but deliver the opposite, salespersons who convince you that you need their product only to leave you with buyer's remorse, or the ubiquitous financial valuation model that lost your money either because it hadn't been properly stressed or back-tested (or worse, had faulty premises to begin with). Your expectation (based on your model) and your experience (reality) failed to match.

As Deirdre McClosky says, metaphor is the most important example of economic rhetoric; economic metaphors are what economists call 'models' (McClosky 1998, 40). The nascent field of Financial Engineering (the author holds a Masters degree in Financial Engineering) which builds valuation and risk models for large investment banks and financial market movers has taken a lot of flak for the Long-Term Capital Management debacle of the late 90s and the more recent crisis in global credit markets–an illustration of what can go wrong when models are not properly calibrated to reality. Nassim Nicholas Taleb's bestseller *The Black Swan* describes at length the anomalies–the "black swans"–that plagued the models created by financial engineers, such as the impact of unusual disastrous events like 9-11.

Anomalies, however, can be a real blessing to those that are inclined to pay sincere attention to them.

> *Scientific development depends in part on a process of non-incremental or revolutionary change. Some revolutions are large, like those associated with the names of Copernicus, Newton, or Darwin, but most are much*

Anomaly

> smaller, like the discovery of oxygen or the planet Uranus. The usual prelude to changes of this sort is, I believed, **the awareness of anomaly**, *of an occurrence or set of occurrences that does not fit existing ways of ordering phenomena.* (Emphasis added) (Kuhn 1977, xvii)

In the gap between the real and its representation lies revolutionary knowledge. Anomalies lie at the heart of what the esteemed philosopher of science Thomas Kuhn calls "scientific revolutions." For instance, Newton's awareness of the falling apple (whether myth or fact) exposed a portion of reality that had been unexplained by the scientific model of the world current then. Newton had found an anomaly, and he revolutionized the way we understand the world. We are still fascinated by the unusual or "anomalous" today. Malcolm Gladwell's *Outliers (2008)* are also anomalies, although of a different kind from Taleb's "black swans" (Taleb 2007). Gladwell spotlights improbable successes, while Taleb's "black swans" are disastrous anomalies.

This book offers my particular perspective as a man with academic degrees in both English and Financial Engineering, with experience as both a professional hypnotist and professional wrestler, and with more than a passing interest in the philosophy of science (believe me, the irony of someone with a resume as anomalous as my own writing a book on anomaly isn't lost on me). My aim is to take both the Kuhnian concept of anomaly and the concept of 'trope' beyond that of the typical philosophy-of-science and rhetoric-of-science audience. Here, you are invited to investigate anomalies critically. Instead of getting caught in analogies and focusing only on what fits in (aka "confirmation bias"), you can make your lives richer and more meaningful by examining and integrating anomalies.

The awareness of anomaly is powerful. The very point of this book is to encourage an open-minded *yet* critical mindset that is aware of the gaps between reality and representation, reviews the phenomenon of anomaly in everyday life, and understands how it can enrich human existence. However, seek not just to be an anomaly hunter but to be an integrator and creator of better paradigms; reconcile the unusual with the expected to create a more accurate representation of reality. While the anomaly hunter can contribute to knowledge through falsification, only one who expands models to integrate anomalies truly extends knowledge.

Anomaly offers the ability to:

- **Discover and implement life-changing knowledge**–Actively seek the unusual and critically integrate into your life the knowledge thus gained.

- **Defend yourself**–When confronted with an uncomfortable feeling that something doesn't fit or sound right, examine the anomaly involved. It could yield important insights and reveal strategies that are meant to manipulate you or hide reality. It's okay to say "I don't know" as long as you then try to figure things out.

- **Express your individuality**–Liberate yourself from analogies, images, and 'identity traps' (an interesting concept I learned about in Harry Browne's classic *How I Found Freedom in an Unfree World* (1998)) that limit you by noting and integrating anomalies within your own identity and relationships. It's your own research program. If well-studied and defended, it can serve as the ultimate expression of your own first-hand understanding of the world. As a hypnotist, I can say that what we call our 'identity' is usually our biggest trance.

Ultimately, this book presents a challenge that was originally thrown down by the famous philosopher Immanuel Kant–*Dare to know!*

Anomaly

A Roadmap To This Book

The first half of this book explores the idea of anomaly and how the gulf between our analogies and reality shape our knowledge and thought. The focus is on the ways in which we come about our knowledge of reality and how that knowledge might be distorted by variances in our expectations and other assorted problems. The lifelong martial artist in me sees this half of *Anomaly* as a lesson in the basics of mental self-defense.

> *Language sets everyone the same traps; it is an immense network of easily accessible wrong turnings. And so we watch one man after another walking down the same paths and we know in advance where he will branch off, where walk straight on without noticing the side turning, etc. etc. What I have to do then is erect signposts at all the junctions where there are wrong turnings so as to help people past the danger points.* (Wittgenstein 1961)

The second half of the book presents marginalized stories (that is, anomalous events) from history and journalism that deserve a second look with the aim of challenging people to integrate these anomalies into their preconceived notions of how the world works. It is an intentionally provocative adventure into the more "fringe" areas of our world that are often promoted by alternative media that, when weighed against the evidence, may have been dismissed prematurely as 'kooky' or 'just another conspiracy theory.'

Please note that within these pages is presented the paradigm I live at the time of writing this book. It presents a snap-shot of where my attention has been focused recently. Please assume *ceteris paribus* and *quod sciam* as implicit throughout because it is quite possible that during my lifetime, my views on some (and perhaps most) may change. As Friedrich Nietzsche says, "Every philosophy is the philosophy of some stage of life."

ANOMALY: REVOLUTIONARY KNOWLEDGE IN EVERYDAY LIFE

My ultimate appeal is for an earnest open-mindedness and critical thinking in our everyday lives.

> *As much as one thinks that businessmen have big egos, these people are often humbled by reminders of the differences between decision and results, between precise models and reality… While many study psychology, mathematics, or evolutionary theory and look for ways to take it to the bank by applying their ideas to business, I suggest the exact opposite: study the intense, uncharted, humbling, uncertainty in the markets as a means to get insights about the nature of randomness that is applicable to psychology, probability, mathematics, decision theory, and even statistical physics.* (Taleb 2007, 268)

CHAPTER 2

BUT IT'S *ANALOGIES* ALL THE WAY DOWN…

A well-known scientist (some say it was Bertrand Russell) once gave a public lecture on astronomy. He described how the earth orbits around the sun and how the sun, in turn, orbits around the center of a vast collection of stars called our galaxy. At the end of the lecture, a little old lady at the back of the room got up and said: 'What you have told us is rubbish. The world is really a flat plate supported on the back of a giant tortoise.' The scientist gave a superior smile before replying, 'What is the tortoise standing on?' 'You're very clever, young man, very clever,' said the old lady. 'But it's turtles all the way down! (Hawking 1988, 1)

The human mind is essentially a black box. That is, we can see the inputs (sensory data) and outputs (behavior, fMRI, etc.), but we don't really know how it works inside. Although we can study the inputs and outputs and try to reconstruct what must have happened in between, often times our only recourse is self-reflection. But if you've ever sat in a barber's chair between two mirrors, you know the strange effect the two mirrors facing each other create. Each mirror will reflect an infinite sequence of progressively smaller reflections of the image on the other mirror. In many ways, something similar happens when you turn your awareness upon itself.

Self-reflection or self-awareness is recursive. Because the object under observation is doing double-duty also as the observer, it creates infinite recursion. The reflexive nature of self-awareness creates a rich, bottomless sensation—you watch your mind, you observe yourself watching your mind, you see yourself observing yourself watching your mind.… The analogy that *self-reflection is like using a telescope to study itself* seems apt to me. It is difficult, imprecise, and ultimately inefficient. The problem is not new as shown by Kant's observation:

> *Through observation and analysis of appearances we penetrate to nature's inner recesses, and no one can say how far this knowledge may in time extend. But with all this knowledge, and even if the whole of nature were revealed to us, we should still never be able to answer those transcendental questions which go beyond nature. The reason of this is that it is not given to us to observe our own mind with any other intuition than that of inner sense; and that it is yet precisely in the mind that the secret of the source of our sensibility is located.* (Kant 1881, 2009)

Another phenomenon that further complicates matters is known to physicists as the 'observer effect'–that the very act of observation affects the phenomenon being observed. So, in addition to recursion, the very act of self-observation could further distort the results of our reflection!

In light of such recursion and distortion, explaining accurately how our mind works is rather difficult, if not impossible. However, this very inability to fully *explain* the workings of our mind offers us a clue to the inextricable relationship between language and thought.

> *I am primarily intrigued by the possibility of learning something, from the study of language, that will bring to light inherent properties of the human mind.* (Chomsky 2006, 90)

The Mighty Pen

That which cannot be expressed in words or symbols lies beyond the reach of the tools of explanation, so, by most definitions, that which is ineffable to the percipient is incomprehensible to the non-percipient. This is why words and symbols play a ubiquitous role in our human experience; from science to art, we are trying to express our understanding of the world to others. Try as we may, the words and sensations used in our explanations of our world are inescapable for as long as we are conscious. No matter how cleverly we try to avoid them, there they are–scribbled in our books; flashing on billboards; broadcasting over the radio and television; dancing as pixels on our computers; popping up in our conversations with families, friends, and strangers; and quite subversively, underlying our most private of thoughts. These explanations can even travel through time and provide tales of the past that come from well before we were born and that will continue to be received by those that are born long after we die...

> Beneath the rule of men entirely great,
> **The pen is mightier than the sword.** Behold
> The arch-enchanters wand!–itself a nothing!–
> But taking sorcery from the master-hand
> To paralyse the Cæsars, and to strike
> The loud earth breathless!–Take away the sword–
> States can be saved without it!
> (Emphasis added) (Bulwer-Lytton 1839, II, ii)

Symbols are indeed mighty; they matter. They can save life; they can kill. Simple words can spark massive epistemological and political revolutions. For instance, the impact of Copernicus's heliocentric model of the cosmos not only reverberated through the scientific circles but also shook the religious spheres of the Renaissance and Reformation years.

At a mundane, unspoken level everyone knows that words and models are powerful and shape our reality, from brilliant psycholinguists like Noam Chomsky, Stephen Pinker, and George Lakoff to your 'Average Joe' on the street. While their models may differ, each with its strengths and paradoxes, incongruities and anoma-

ls and words, whether clear or muddy, bear serious
Gandhi's model and his words gained India its in-
le Hitler's led the world into one of its worst wars.

> *In the beginning, the Word existed. The Word was with
> God, and the Word was God.* (John 1:1)

In *Genesis*, Adam became the master of the world when he took control by assigning names "to all cattle, and to the fowl of the air, and to every beast of the field." In the ancient world, naming was associated with possession and dominion. Maybe, language is what has alienated us from the natural world, but then maybe, it has really given us greater integration or 'oneness' with it by giving us the tools to manipulate and think abstractly about it.

> *Noam Chomsky and others maintain that that which is
> essentially human is our innate linguistic ability, a sort
> of 'primal grammar' that all languages would share at a
> sufficiently deep level.* (D. Hofstadter 1982, 106)

The Conflict

The words of the First Amendment to the Constitution of the United States insure the freedom of speech, while much of the nation's National Security budget goes toward cryptology, the monitoring of the speech of American citizens. The tension between the two illustrates the power that language possesses to not only model our reality, but to shape it too.

> *Why shouldn't we quarrel about a word? What is the
> good of words if they aren't important enough to quarrel
> over? Why do we choose one word more than another if
> there isn't any difference between them? If you called a
> woman a chimpanzee instead of an angel, wouldn't there
> be a quarrel about a word? If you're not going to argue*

> about words, what are you going to argue about? Are you going to convey your meaning to me by moving your ears?
> (Chesterton 2004)

Law is written elaborately to avoid ambiguities of any kind so that everyone understands it the same way; however, we see laws being interpreted and re-interpreted–an illustration of the gap between language and reality. However much we try to be unambiguous, we do not succeed completely because the reality we try to capture is more complex than our tools. This is why mathematics is the *language du jour* of science, to help avoid these ambiguities with specific mathematical symbols. Yet, despite the best efforts of science, the slippage between these symbols and reality occur.

Many philosophers have noticed this slippage between language, thoughts, and reality. For the Ancient Greeks, thinking was tightly bound to language. In fact, the Greek word for reason *logos* also refers to speech. For the naturalists in the tradition of Plato, the sounds of words came directly from and were intrinsic to those objects that they actually referred to. However, as David Crystal points out, the infrequent instances of onomatopoeia show the naturalists' case to be a weak one (Crystal 2005). For the conventionalists in the tradition of Aristotle, the relationship between words and the things they referred to were arbitrary. All words, including the onomatopoeic, were symbols that stood for concepts/idea of something in reality.

> *Words are symbols created to represent reality. A term is a concept communicated through a symbol. Once words are used to communicate a concept of reality, they become terms.*
>
> *Communication is dynamic; it is the conveying of an idea from one mind to another through a material medium, words or other symbols. If the listener or reader receives through language precisely the ideas put into it by the speaker or writer, these two have 'come to terms'–the idea*

> has passed successfully, clearly, from the giver to the receiver, from one end or term of the line communication to the other.
>
> A term differs from a concept only in this: a term is an idea in transit, hence is dynamic, an **ens communicationis**; the concept is an idea representing reality, an **ens mentis**. A concept is a potential term; it becomes an actual term when it is communicated through a symbol. Hence a term is the meaning, the form, the logical content, of words. Words are therefore the symbols, the means by which terms are conveyed from mind to mind. (Joseph 2002, 71)

Words, in short, are doubly removed representations of reality and, therefore, potential Trojan Horses.

How Much Can We Know?

> When the human mind, with its limited powers, attempts to mirror in itself the rich life of the world, of which it itself is only a small part, and which it can never hope to exhaust, it has every reason for proceeding economically. ...In reality, the law (Laws of Nature) always contains less than the fact itself, because it does not reproduce the fact as a whole but only in that aspect of it which is important for us, the rest being intentionally or from necessity omitted. (Mach 1968)

Philosophers like Aristotle and Kant and today's cognitive neuroscientists have all sought the limits of human knowledge. In an attempt to make the problem manageable, it's often bifurcated. Philosophy has had a tradition of separating knowledge into two categories, the *a priori* and *a posteriori*. *A priori* knowledge is knowledge independent of experience. Perhaps, a better way to think about *a priori* knowledge is as knowledge already 'built-in.' Chomsky's concept of Universal Grammar, the idea that everyone is born with some innate proclivity or principles for language, is a sort of example of this built-in knowledge. *A priori* statements, such as 'all children are young' or 'all bachelors are unmarried,' are always true because the concepts 'young' and 'unmarried' are built into the meanings of 'children' and 'bachelors' respectively and the other way round. *A posteriori* knowledge is knowledge derived solely from experience and is falsifiable. For example, 'some bachelors are young.'

Scottish Enlightenment philosopher David Hume, whom Kant aptly dubbed the "geographer of human reason," has a similar division—between the 'relations of ideas' and 'matters of fact.' For Hume, mathematics and logic, the manipulations of symbols, really had nothing to do with reality and, as such, were simply tautological (Hume 1748). In a similar vein as Hume, the influential scientist and philosopher Ernst Mach (the speed of sound is named after him) held that the Laws of Nature are but the sum of sensory experience and nothing more (Mach 1968). Influenced heavily by Mach, lesser-known German philosopher Fritz Mauthner also rigorously explored the philosophy of language and sought to extend the epistemological critique of language made by Mach:

> *Language never coincides with nature, even where real or approximate laws have been found: in mathematics, in mechanics.* (Mauthner 1910-11)

> *If thought and speech were one, said Mauthner, language was not an instrument of thought but* **"nothing other than its use (Gebrauch). Language is the use of language."** *No knowledge was possible apart from language, yet language was inadequate to the task. Sensation were contingent, the meaning people gave them stable; the same words were used to describe invariably different sensations. This made it impossible for language to describe*

> *the world with any precision, and rendered all language metaphorical.* (Hacohen 2002)

While not well known today, Mauthner was one of the few philosophers that Ludwig Wittgenstein took to task, by name, in his book *Tractatus Logico-Philosophicus*, arguably one of the most influential philosophical writings of the twentieth century. Ironically, eventually Wittgenstein and Mauthner held some nearly congruent beliefs about language and the use of language:

> *For a large class of cases – though not for all – in which we employ the word 'meaning' it can be defined thus: the meaning of a word is its use in the language.* (Wittgenstein, PI 43 2009)

Modern semanticists too see it the same way. They look at the sense of the word; that is to say, its *use* in language, not necessarily what the word *refers* to. Again, we are faced with the gap between what is knowable because language can express it and reality.

> *Philosophy is written in this enormous book which is continually open before our eyes, but it cannot be understood unless one first understands the language and recognizes the characters with which it is written. It is written in a mathematical language, and its characters are triangles, circles, and other geometric figures. Without knowledge of this medium it is impossible to understand a single word of it; without this knowledge it is like wandering hopelessly through a dark labyrinth.* (Galilei 1864)

While Mach, Mauthner, and Wittgenstein saw this gap between language and reality as damning for science (ultimately relegating one to silence), the philosopher of science Thomas Kuhn saw this gap as an opportunity. He saw this gap as the very catalyst for the great revolutions in knowledge. He called this gap an anomaly (Kuhn 1977).

But It's Analogies All The Way Down...

As Kuhn saw it, 'normal' science is driven by adherence to a particular model, world-view, or paradigm, which provides scientists questions to answer and the tools to answer them with. For example, in linguistics, words or morphemes are re-formed or created based on the models of existing grammatical patterns in a language, that is, through analogy. This leads to greater regularity in paradigms. Mathematics too is a system of analogies. Take, for example, Einstein's famous equation: $E=mc^2$. It is an analogy that uses the concepts of mass and the speed of light to explain the concept of energy.

Occasionally, however, a crisis arises when paradoxes begin to undermine the current explanatory model. These paradoxes or gaps between what is expected by the paradigm and what really happens are called anomalies. Revolutionary science resolves the crisis by convincingly integrating the anomalies into a newly accepted paradigm. In the Kuhnian sense, everything that we've been talking about up to this point–the slippage between a word/model and the corresponding reality–offers not just crisis, but also an opportunity for revolutionary knowledge!

On December 21, 1954, a worldwide flood was predicted to bring an end to the world. The doomsday UFO cult that made this prediction expected aliens to rescue them at 12 midnight, the day of the flood. As you guessed, the Armageddon never came, nor did the UFO. The book *When Prophecy Fails* chronicles how the cult adapted its expectations in the face of reality. As they sat and prayed that night, they noticed a glaring anomaly: the cataclysmic flood, as well as the promised UFO, failed to materialize. The world was still alive and they were still in it. One of the book's authors, Leon Festinger, coined the term 'cognitive dissonance' to describe the mental state of the cult's true believers when their prophecy failed to come true. Instead of abandoning their model in the face of the anomaly, they resolved their cognitive dissonance by giving their God a merciful face. Their God spared the planet because of their prayers (Festinger 1956).

Oftentimes, anomalies are marginalized because of the inconveniences and psychological turmoil they can cause. Very few people are comfortable with that annoying feeling of 'cognitive dissonance' that accompanies anomalies, and revising the dominant paradigms of our personal lives can sometimes be painfully difficult. Most commonly accepted paradigms, whether scientific or not, tend to offer their user some combination of prediction, control, falsification (what I call anomaly hunting), and/or explanatory power.

For example, classical mechanics (the physics equations you learned in high school) presents excellent examples of models that offer predictive ability, control over things, falsifiability, and great explanatory power (not at the sub-atomic level of course). The UFO doomsday cult had a model too; unfortunately, it wasn't very accurate in its prediction and was very easily falsified.

> *When you yourself have a difficult decision to make involving unknown quantities in the future, you do go in for a form of simulation. You **imagine** what would happen if you did each of the alternatives open to you. You set up a model in your head, not of everything in the world, but of the restricted set of entities which you think may be relevant. You may see them vividly in your mind's eye, or you may see and manipulate stylized abstractions of them. In either case it is unlikely that somewhere laid out in your brain is an actual spatial model of the events you are imagining. But, just as in the computer, the details of how your brain represents its model of the world are less important than the fact that it is able to use it to predict possible events. Survival machines which can simulate the future are one jump ahead of survival machines who can only learn on the basis of overt trial and error. The trouble with trial and error is that it is often fatal. Simulation is both safer and faster.* (Dawkins 1976)

But It's Analogies All The Way Down...

As Alfred Korzybski says, "Man's achievements rest upon the use of symbols." (Korzybski 1958, 76). Symbols, analogies, models, terms, and metaphors shape the way we think about our world and ourselves. They must be treated discretely and with respect since these models ultimately form the foundation of our knowledge.

Schemes and Tropes

Before moving on to the question of how we develop or acquire these models, let's take a quick peek at analogies and metaphors in the context of classical rhetoric. These more amphibolous models and structures are often called 'trope' and 'schemes.' While trope is an unconventional use of a word, a scheme is an unconventional word order, both used to create an impact on the listener/reader. If analogy is the lifeblood of thought, ambiguous language is the lifeblood of trope. Used in poetry or prose for beauty, they truly enrich our lives, but in the hands of a con-artist, these become tools for manipulation. In his dystopian novel *1984*, Orwell called it Newspeak.

> *In classical rhetoric, the tropes and schemes fall under the canon of style. These stylistic features certainly do add spice to writing and speaking. And they are commonly thought to be persuasive because they dress up otherwise mundane language; the idea being that we are persuaded by the imagery and artistry because we find it entertaining. There is much more to tropes and schemes than surface considerations. Indeed, politicians and pundits use these language forms to create specific social and political effects by playing on our emotions.* (Cline 2006)

While it's analogies all the way down, awareness of the models that we've adopted, either consciously or unconsciously, and that are reflected in our everyday thoughts, words, emotions, and actions will empower us not only to withstand manipulations but also to enrich our lives by seeking anomalies and integrating them so we live closer to reality. The earnest anomalist concerns themselves with the differences that will make a difference–between protoscience and pseudoscience, between Type I errors and Type II errors (discussed later), and ultimately, between reality and analogy.

CHAPTER 3

FIRST-HAND, IN-BUILT, AND EPISTEMIC HAND-ME-DOWNS

How do we *know* reality? How do we *know* that we aren't, say, in 'The Matrix'? French intellectual Jean Baudrillard claimed to have had difficulty differentiating between cable news programming, video game simulations, and the military-media narrative presented as the Persian Gulf War (Baudrillard 1995). Baudrillard's claim may seem a little over-the-top, but it's not totally off the mark. This is one of the basic epistemological questions that philosophers down the ages have attempted to answer, and because we don't literally have the 'red pill' that will take us out of the Matrix, we can only say with Descartes "I think, therefore I am." (Descartes 1996). Fortunately, the very act of 'doubting existence' implies that there is an existence to doubt. And given that *something* exists, we can do our very best to simply make sense of it.

Our senses provide us with multiple access points to reality. A child seems to know ahead of experience, for example, that a wall is solid and will not allow it passage. (You'll rarely see a child, however young, walk into a wall, unless deliberately, at play.) Experience teaches the child that hot things can hurt, and of course, we have all heard stories or read books or watched TV and gained knowledge about reality. In this way, our simulations are yet another form of analogy, which allows us to predict how a model of reality will work.

> *Through the technique of simulation, model battles can be won or lost, simulated airliners fly or crash, economic policies lead to prosperity or ruin. In each case the whole process goes on inside the computer in a tiny fraction of*

> the time it would take in real life. Of course there are good models of the world and bad ones, and even the good ones are only approximations. No amount of simulation can predict exactly what will happen in reality, but a good simulation is enormously preferable to blind trial and error. Simulation could be called vicarious trial and error, a term unfortunately preempted long ago by rat psychologists. (Dawkins 1976)

First-hand and Inbuilt Knowledge

Assuming that we can move beyond the assumption that we are all just 'brains-in-vats' stuck somewhere in a basement laboratory of Wilder Penfield, it seems reasonable to suggest that our sense organs and nervous system gather and organize data about the world of which we are a part. This feed-forward and feed-back system shapes the world we know first-hand. For example, you know (as well as you know anything) that you are indeed reading these words right now, you know where you are right now, you know what your clothes feel like against your skin, and so on. This first-hand experience is as certain as things can get.

However, first-hand experience is only one part of the story of how we *know*. Modern research has shown that we are, in fact, born with *a priori* knowledge. Our minds are not, as was earlier believed, *tabula rasa* or 'a blank slate'. Developmental psychology has shown that infants are born with the ability to recognize continuity, causality, and form. In my own experience, I saw how my infant daughter instinctively 'knew' how to cry when she needed something, and there was no learning curve involved, as best I could tell.

A more famous example of this 'baked-in' knowledge comes from the longitudinal identical twin study done at the University of Minnesota, which suggests that many of our 'freely' chosen biases are, in fact, 'built-in.' In the study, the researchers followed

First-Hand, In-Built, And Epistemic Hand-Me-Downs

identical twins separated at birth over several years. Because they were genetic clones, that is to say, they shared precisely the same genetic code, identical twins separated at birth presented an outstanding opportunity to grasp the impact of nature vs. nurture (since the nature-side is controlled for).

The findings were truly amazing. In *Entwined Lives*, Nancy Segal describes how, without foreknowledge of each other, these twins named their pets and children the same names, married the same number of times to people *with the same name*, drove the same car, vacationed at the same location, had the same preferences in food, beverages, and cigarettes, had the same occupations, and so on (Segal 2000). These are truly uncanny circumstances and, given the nature of joint probabilities, too highly improbable to be fairly called 'coincidences.' The apparent choices of these identical twins seemed driven more by nature than nurture and questioned the bounds of free will. While this doesn't bode well for the advocates of free will, the Minnesota Twins Study does add another side to our 'Archimedesian polygon'[1] of knowledge, bringing us even closer to understanding the realities of our own thought and decision processes.

To take our understanding of in-built knowledge further, scientist and consciousness researcher Benjamin Libet designed clever experiments that effectively timed both the conscious decisions of participants to act and the brain activity associated with the physical initiation of behavior and then compared the times. The results suggest that just *before* the participant made a conscious decision to act, their actions were actually already happening. The implication is that while we may not consciously be the prime-movers of our own actions, it seems that there may be some ability to stop or steer these unconscious actions once they are already in motion (Libet

---

[1] As a way to calculate π, Archimedes used a polygon. A polygon would begin to approximate a perfect circle as the number of sides of the n-sided polygon increased toward infinity. However, no matter how many sides were added, the polygon would still only be an approximation of a perfect circle, but good enough for practical purposes.

2004).

The results of both the Minnesota Twins study and Benjamin Libet's research are thorny anomalies because they imply that we really may not have much of what is called free will (although, some of the Minnesota study's participants parted their hair differently and had other notable differences). It seems to tell us that somewhere between our genetics, the laws of basic mechanics, our creative ability, and our ability to avoid false alternatives lies some sort of *Bounded Choice Set* that we call free will. It seems that we are in a restaurant with a limited menu–the predetermined. However, given the menu, we can choose what to eat–our bounded choice set or free will. The steering mechanism suggested by Libet's study that stops or steers unconscious actions may be what we call thought or the ability to reason, which aims to help us make the best choices given what we know *a priori* and *a posteriori* first-hand, despite the evidence that our will isn't entirely our own. So how do we use reason to steer us toward those signals that exhibit the highest fidelity to reality?

Epistemic Hand-Me-Downs

> *Humans are no longer dependent for information upon direct experience alone. Instead of exploring the false trails others explored and repeating their errors, they can go on from where others left off. Language makes progress possible.* (Hayakawa cited in Morville (2005))

Myths, legends, folklore, and various oral traditions have brought us knowledge acquired by our very remote ancestors. Add to that the immense amount of written material humankind has produced, and we have a massive repertoire of knowledge available to us: the first-hand experience of others–the ancients, as well as contemporaries–made available through language.

A personal aside: I love reading and word craft. This made me pursue a bachelor's degree in English, although I later moved on

First-Hand, In-Built, And Epistemic Hand-Me-Downs

to become a Financial Engineer after finding difficulty procuring a well-paying job with an English degree. In hindsight, however, I see something particularly special that happened in those four years of full-time study of the English language as an undergraduate in Boulder that I didn't appreciate until recently. The chance reading of Ayn Rand's *Fountainhead*, at the behest of a college roommate, opened up a whole new world of ideas to me. In addition to the typical critical theory texts of Ferdinand de Saussure (Marxist), Jacques Derrida (deconstructionist), Jacques Lacan (Freudian), and so on, presented in the typical undergraduate English major curriculum, I also began to include the texts of Karl Popper, Robert Nozick, Thomas Kuhn/Michael Polanyi, and many, many others. Eventually, my reading list broadened to include economics and even mathematics.

While working in the shipping department of a libertarian direct-mail bookstore, I began enjoying Austrian economists like Friedrich von Hayek, Ludwig von Mises, Murray Rothbard, and Israel Kirzner. Their explanations of economic concepts in plain English allowed me to extrapolate the meaning of what other economists, who employed more statistical explanations in their presentation, were discussing. Though, at one time, I had a decent-sized phobia of mathematics and science (another reason for opting English), I could settle upon a very mathematically oriented Master of Science of Financial Engineering program because my reading helped me understand mathematics as merely another language, a tool to help communicate and understand reality. Since then, I've worked professionally as a marketing statistician at a direct-mail database cooperative and as the modeling manager for the largest (at the time anyway) originator of reverse mortgages in the country.

As I read and understood others' models of reality, I gained a greater command over language, which meant more clarity of thought and a richer ability to understand the world in which I lived. I began to understand the appeal of speaking multiple languages (although I haven't had much time to devote to it). It also made me realize that to control one's thought is to control one's use of terminology and reasoning. The master of mind-control (over one's self or others) is nothing more than the masterful rhetorician.

Tacit and Explicit Knowledge

When we visit a new restaurant, we do not have first-hand knowledge of the food there. We rely on the menu–the menu-maker's first-hand knowledge made explicit. Chemist/philosopher Michael Polanyi calls first-hand knowledge 'tacit knowledge,' and epistemological hand-me-downs (the menu) he calls 'explicit knowledge' (Polyani 1962).

Explicit knowledge is twice removed from reality. It's reality as understood by the percipient and put into words. Going back to our menu metaphor, we do not consume the menu or the descriptions on the menu. The menu is the menu-maker's attempt to describe in words the food available at the restaurant. Our satisfaction will depend to a great extent on how our first-hand experience of the food matches the expectations set by the descriptions on the menu. Our expectations are Kuhnian paradigms of sorts (Kuhn 1977). When our expectations are met, we experience satisfaction and proceed unaffected. However, given uncertainty, when there is an anomaly (that is, a deviation from our expectation), we experience either disappointment or pleasure and (hopefully) revise our expectations. Finance attempts to quantify this phenomenon of deviation around expectation using the statistical concept of a standard deviation (which requires normally distributed data, or a bell curve). Unfortunately, as Benoit Mandelbrot has shown, it seems that seldom is a bell-shaped curve an appropriate metaphor for how chance really works (Taleb 2007). The tails are much fatter. Randomness and unpredictable outliers are out there, waiting to sucker punch you. We'll come back to this in Chapter 13.

Both tacit and explicit knowledge can vary from person to person. The same curry could taste hot to one and mild to another and, accordingly, its description and reader-understanding. Both first-hand knowledge and second-hand knowledge should come with user-manuals but don't, so we all learn to craft our own (the scope and accuracy of which varies greatly from person to person). As my exposure increased to the descriptions, models, and analogies of great thinkers, I found myself with a larger vocabulary to describe reality and a greater ability to understand and sympathize with what's happening in another's world.

However, the very best and honest attempt to describe and explain reality in ordinary language or even by means of the terms of science and mathematics is still only the very best attempt. It's like adding sides to Archimedesian polygon (see Footnote 1). Language, in short, is a rather less than perfect tool for understanding reality, but it's the best we have.

But that is not the end of our woes; humans make mistakes, fabricate, and exaggerate.

> If we would speak of things as they are, we must allow that all the art of rhetoric, besides order and clearness; **all the artificial and figurative application of words eloquence hath invented, are for nothing else but to insinuate wrong ideas, move the passions, and thereby mislead the judgment; and so indeed are perfect cheats**; and therefore, however laudable or allowable oratory may render them in harangues and popular addresses, they are certainly, in all discourses that pretend to inform or instruct, wholly to be avoided; and where truth and knowledge are concerned, cannot but be thought a great fault, either of the language or person that makes use of them. (Emphasis added) (Locke 1997, Bk. III, 307)

As explicit knowledge is the result of the epistemological (analogy building) efforts of many human beings, there is the very real risk that we may be seeing the world through the colored lenses of many mistakes, fabrications, and exaggerations that have been institutionalized and accepted as 'knowledge' without critical review. That is, we may be living poorly calibrated analogies, and only attention to anomalies can anchor us to reality. Both lies and truth present models of reality, and the anomalies presented by *both* offer us the opportunity to create a richer life.

Language and Thought

What is the impact of language on thought and, therefore, knowledge? For Polanyi, it's everything–the scientist's language ('vocabulary and structure') constrains the questions that can be asked and the questions asked determine the answers that can be discovered (Polyani 1962). Linguists Sapir and Whorf are perhaps the most famous promoters of the idea that your language limits the extent to which you can think (or not think) about a particular subject, although Orwell's Newspeak in *1984*, where the control of language is equivalent to the control of thought, actually has been more popularly influential.

> *How could you have a slogan like "freedom is slavery" when the concept of freedom has been abolished? The whole climate of thought will be different. In fact there will be no thought, as we understand it now. Orthodoxy means not thinking–not needing to think. Orthodoxy is unconsciousness.* (Orwell 1950)

There are many critics of the Sapir-Whorf hypothesis, most notable among them Stephen Pinker and Noam Chomsky. Although they emphasize that thought precedes language, they do not deny that thought and language are entwined. Cognitive linguists (like Lakoff), philosophers (like Daniel Dennett), and epistemic rhetoricians (like Deirdre McClosky and Alan G. Gross) have raised tough arguments that demonstrate the metaphoric nature of language and

First-Hand, In-Built, And Epistemic Hand-Me-Downs

lay emphasis on the central role of discourse in the creation and validation of knowledge.

> *Without a contradiction between language and reality there is no mobility of concepts, no mobility of signs, and the relationship between concepts and signs become automatized. Activity comes to a halt, and the awareness of reality dies out.* (Jakobson 1933)

A Note on the Mechanics of Communication

Communication is expressing our thoughts, and it is successful when our thoughts have reached and been interpreted by the listener or reader as we intended. Claude Shannon (no relation to your author), the Father of Information Theory, along with his colleague Warren Weaver put forth a model of communication that, by removing meaning from communication, was able to transmit pure units of information known as binary digits or 'bits' from one place to another. Simply stated, Shannon's model of communication presented information as a discrete unit transmitted from a source, through a channel, into a receiver. By focusing upon the encoding of transmitted information and the decoding of received information, Shannon's mathematical model of communication allowed for the recognition of the effects that **noise** had on the information being communicated.

Their original model had 6 basic components: the source, the transmitter, the channel, the receiver, the destination, and noise. For example, take our communication via this book. I am the source (with my message about anomaly), this book is the transmitter (encodes my message into printed words), the wholesale market for books is the channel, the retail store is the receiver, and you, my noble reader, are the destination. Noise, in this example, could be any ripped, missing, or marked upon pages that interfere with your receiving the full message I've sent. In *Metaphors We Live By*, cognitive linguists George Lakoff and Mark Johnson reiterate Michael Reddy's observation that Claude Shannon's 'conduit metaphor' of communication is ubiquitous. They describe it thus, "The speaker puts ideas (objects) into words (containers) and sends them (along a conduit) to a hearer who takes the idea/objects out of the word/containers" (Lakoff and Johnson 1980, 10).

Roman Jakobson, a brilliant Russian émigré, refined Shannon's 'conduit' model by reintroducing the importance of human meaning. To bring meaning, reason, and motive back into the model, he added three more elements: context, message, and code. In the context of this book, the English language we're using is the code, the message is the word/phrase being transmitted, and the context is the referent in reality.

Knowing reality is an ongoing process, and central to the acquisition of new knowledge is the anomaly, or the gap between any particular model and reality. To enter this process effectively, we need tools of reasoning, and that's what we move on to next.

CHAPTER 4

REASONING: OUR TOOL FOR LEARNING

The most formidable weapon against errors of every kind is reason. (Paine 1975)

How best do we deliberately seek out anomalies in our lives and use them to refine our models of the world in a way that makes our understandings richer and more accurate? The tool at our disposal are our symbols, and given the central role of language in our thinking, model building, and anomaly seeking, it is time to resurrect a curriculum that has sadly become somewhat of an anomaly itself: *the trivium*.

In the Liberal Arts, there are seven branches of knowledge upon which our knowledge of the world is built. The first three, also known as the *trivium*, are made up of the three Rs–not of reading, 'riting, and 'rithmetic, but of 'riting (grammar), reckoning (logic), and rhetoric. The *trivium* was preparatory for the *quadrivium*, the remaining four branches that dealt with matter–arithmetic, music, geometry, and astronomy (with the first two dealing with discrete quantities and the latter pair dealing with continuous quantity).

Made up of the three arts of language, the *trivium* lays the basis for understanding and building models of the world and is important to our quest to understand better the role of anomaly in our lives.

> *Logic is the art of thinking; grammar, the art of inventing symbols and combining them to express thought; and rhetoric, the art of communicating thought from one mind to another, the adaptation of language to circumstance.* (Joseph 2002)

Anomaly: Revolutionary Knowledge in Everyday Life

(I highly recommend Sister Miriam Joseph's book for its straightforward presentation. Much of what follows here owes greatly to her explanations, and as anyone who has read her book will attest, my debt to her writing will become clearer as we proceed.)

Grammar: The Thing-As-It-Is-Symbolized

Grammar is the most basic of the Liberal Arts. It's the craft of inventing and combining symbols (words or terms) to form correct sentences that convey ideas clearly. The foundation upon which all of this rests is made up of 'terms'–words that carry concepts. Terms are analogies that stand for some aspect of reality. When you 'come to terms' with someone or something, it means that you are in agreement or acceptance of the reality represented through your terms.

Terms exhibit qualities known as extension and intension. They are inversely related–the more extension a term has, the less intension it has (and vice versa). The extension of a term is the total set of objects it can refer to. For instance, the extension of the term 'swan' is all swans–white or black. While extension is a measure of abstractness and ambiguity, intension is the definition or the meaning of a term. For example, the intension of the term 'swan' would comprise all qualities that distinguish it from other birds, or in fact, everything else. The more discrete and specific a term, the fewer the things it can possibly refer to. While disciplines, such as science and law, require terms with more intension (less extension) for precision and clarity in communication, religious prophecy and psychic readings (not genuine psi phenomena in Radin's sense, but the 'cold readings' of psychics) often use terms with large amounts of extension leaving room for ambiguity. However, if terms do not have adequate extension, the result could be cognitive dissonance and anomalies, such as with the case of the UFO cult (who noticed an anomaly when the world didn't end according to the schedule set by their model of the world) or the proponents of the pre-heliocentric model of the solar system (as when Copernicus, and later Galileo, presented strong evidence that proved the geocentric model to be very wrong). In this way, the concepts of intension and extension are parallel to Frege's concepts of *sense* and *reference* (for example, the names *Borat, Ali G, Brüno,* and *Sasha Baron Cohen* all refer to the same person, but they present that person in meaningfully different ways. In other words, the four names or references (extension or denotation) each have different senses (intension or connotation).

Grammar categorizes words/terms into lexical categories that behave in a particular way when put in a sentence. Grammar also describes how these combine to form correct sentences carrying meaningful propositions. Correct propositions then allow us to communicate in a logical, non-contradictory fashion. Grammar enables accurate expression of thoughts, models, and analogies through correct, unambiguous sentences and is, therefore, the most basic of liberal arts.

Logic: The Thing-As-It-Is-Known

Logic is the mechanics of reasoning or *good* thinking. Once we are adept at the craft of inventing and combining terms to convey ideas clearly, we need to be able to build our argument or model. Until the arrival of the revolutionary paradigms of Frege and Peirce, Aristotle's logic had remained largely unchallenged. Its influence on the history of Western thought is unparalleled. Now, there are several systems of logic–sentential logic, predicate logic, and mathematical logic to name a few. As it is not within the scope of this book to discuss these systems and you've easy access online to detailed descriptions, we'll only try to get a feel of how logic works by discussing deductive and inductive logic.

Aristotle's famous syllogisms are a form of argument (a formal relationship of propositions) where two propositions (aka 'premises') share a common term which leads to a third proposition called a 'conclusion.' The 'conclusion' is argued to be true because of the way the 'premises' are related. Here is an example:

> *All swans are white. The Australian bird is a swan. Therefore, the Australian bird is white.*

The term 'swan' has a greater extension than 'the Australian bird' because the second proposition includes 'the Australian bird' in the extension of the term 'swan,' and therefore, what is true of a 'swan' is true of 'the Australian bird.' The way this argument is built is called deductive logic. But what happens to this deductively valid argument when an anomalous black swan comes along? While the original syllogism may still be 'valid' because it is built per the rules of logic, ignoring the black swan will result in a poor decision in reality. As software engineers are fond of saying, garbage in, garbage out.

Another form of logic Aristotle talks about is inductive logic where you build from the general to the particular. Induction lies at the nexus of reality and deductive reasoning. Human beings love to make sense out of chaos. We are always on the lookout for patterns. When we observe patterns of repeated experience, we build analogies. For example, so long as we have observed only white swans, whiteness is one of the defining characteristics of the term 'swan,' and renders the proposition 'all swans are white' true. Induction allows us to reason in the face of uncertain outcomes, that is, having observed *n* number of white swans does not guarantee that the next swan you see will be white. Inductive reasoning also is where a lot of the slippage between our expectations and reality occur.

> ...in Peirce's phrase, inductions are **ampliative**. Induction can amplify and generalize our experience, broaden and deepen our empirical knowledge. Deduction on the other hand is **explicative**. Deduction orders and rearranges our knowledge without adding to its content.

Reasoning: Our Tool For Learning

Of course, the contingent power of induction brings with it the risk of error. Even the best inductive methods applied to all available evidence may get it wrong; good inductions may lead from true premises to false conclusions. (Vickers 2009)

Both these types of logic have their value. While induction builds analogies by observing reality, deduction allows new insights to emerge by arranging and rearranging propositions already held true. However, both have their dangers. As said earlier, if any of the premises held true is not true, the conclusion drawn will not be true in the real world even though the argument may remain 'valid.' In other words, have a healthy skepticism of authority. With inductive thinking, it is important not to be seduced by statistics, especially with regards to prediction. As Dan Borge says:

Everyone knows that the past is, at best, an imperfect guide to the future. However, risk experts, like experts in other fields, are prone to fall in love with their tools and this love can lead to severe myopia. Experts can be tempted to define problems in ways that fit their tools rather than ways that fit the actual situation. Statistics is a favorite tool of the risk expert and the past is much more accessible to statistics than the future. Only the past has the data points that statistics craves. So the unwary risk expert may exaggerate the importance of the historical data that allow him to use his favorite tools and to arrive at a definite solution, even if it is the solution to the wrong problem. (Borge 2001, 55)

A Note on Science, Falsification, and Skepticism

British empiricist David Hume's 'problem of induction'—does inductive reasoning lead to truth?—has troubled thinkers for centuries. No amount of confirmatory observation can finally prove as true an inductively reached conclusion, such as a law of physics.

> *All the objects of human reason or enquiry may naturally be divided into two kinds, to wit, Relations of Ideas, and Matters of Fact. Of the first kind are the sciences of Geometry, Algebra, and Arithmetic ...[which are] discoverable by the mere operation of thought...Matters of fact, which are the second object of human reason, are not ascertained in the same manner; nor is our evidence of their truth, however great, of a like nature with the foregoing.* (Hume 1748)

Two centuries later, Sir Karl Raimund Popper promoted the revolutionary proposal that the goal of science is not to confirm theories, but rather to falsify them. No amount of verifying can establish a theory about reality as true. Science should, therefore, attempt to falsify–hunt for anomalies–and theories that are not falsifiable are not scientific. In the words of Alan G Gross, "Falsifiability is prediction turned on its head" (Gross 1990, 40). To test a theory or model is to test the predictions it makes about how reality will function. Failed predictions reveal the bounds of the theory. Under this interpretation of science and the scientific, the body of human knowledge grows by error-elimination much like evolution is driven by natural selection. In short, the scientific method is to look for anomalies and explain and integrate them into our analogies making them fit reality better.

When we survey scientific literature, we observe the themes of predictability, control, falsifiability, and explanatory power recurring often. It is these variables, and the respective weight of each, that determine how scientific a statement is–the more scientific a statement, the higher its score on all variables. Newton's equations for classical physics scored high across the board (at least, until the anomalies present at the sub-atomic level demanded changes). The models of the psychologist, astrologer, and economist often score high on explanatory power but low on predictability and control. Social scientist Hayek, as well as Popper, implicitly warns us against the tendency to grant excessive weight to explanatory power (for example, as in scientism), without seeking high scores for other equally important variables.

This isn't to suggest, however, that we throw the baby out with the bathwater; instead, we need to approach science critically, albeit with an open-mind. This brings us to 'scientism': a "pejorative term for the belief that the methods of natural science, or the categories and things recognized in natural science, form the only proper elements in any philosophical or other inquiry" (Oxford English Dictionary). It looks like science but turns science and accepted methods of science into a religion by taking them beyond the purview of falsification. For science to remain science, it must be falsifiable; it must have adequate intension.

Famed skeptic Michael Shermer has used the term *scientism* in an article to describe "a scientific worldview that encompasses natural explanations for all phenomena, eschews supernatural and paranormal speculations, and embraces empiricism and reason as the twin pillars of a philosophy of life appropriate for an Age of Science" (Shermer 2002). In this sense, scientism is no different from naturalism or simply science. However, this is trope. Scientism, as Hayek and Popper use the term and as it was originally used, gives undue authority to 'scientific' claims, which then ultimately undermines what science stands for. Scientific claims are human endeavors and, as such, they are subject to error, ambiguity, and/or deception. They warrant scrutiny. However, Shermer ends his article with words that do capture the crucial role of language in the scientific endeavor that makes it storytelling of a specific kind:

> *Because of language we are also storytelling, mythmaking primates, with scientism as the foundational stratum of our story and scientists as the premier mythmakers of our time.* (Shermer 2002)

At the other extreme of scientism sits skepticism. A skeptic (in the tradition of Phyrro, at least) does not take a position. It's like agnosticism, which makes no claims either way about the existence of God. An agnostic claims that there is not enough evidence to prove the existence of God and that the absence of proof isn't the same as proof of absence.

> *Both critics and proponents need to learn to think of adjudication in science as more like that found in the law courts, imperfect and with varying degrees of proof and evidence. Absolute truth, like absolute justice, is seldom obtainable. We can only do our best to approximate them.* (Truzzi 1987)

Errors, Cognitive Illusions, and Informal Fallacies

To its believers, the UFO cult's theory about the world could seem high on explanatory power and predictability as it predicted doomsday and was able to explain why the prediction did not materialize. But rather obviously, their theory or model is far from reality. When building models, it is not uncommon for errors and fallacies to creep into our reasoning, which take us away from reality. Before moving away from the branch of logic, we'll take a quick peek at some of these errors and fallacies. (As with logical systems, it is beyond the scope of this book to deal comprehensively with the potential errors, fallacies, and illusions in reasoning.)

Pattern Finding

Our ancestors attributed random lightning and thunder to the anger of the gods (Zeus for Greeks, Indra for Hindus, Thor in Norse mythology, etc.). As children, many of us looked at the moon and saw a man's face. Finding a pattern where none exists is common. It's the projection of meaning upon meaningless noise. Though misleading, believing a false pattern (also called a false positive or

Type I error) could be less fatal than believing that no pattern exists when in reality it does (a false negative or Type II error).

> *Harvard University biologist Kevin R. Foster and University of Helsinki biologist Hanna Kokko...demonstrate that whenever the cost of believing a false pattern is real is less than the cost of not believing a real pattern, natural selection will favor patternicity. ...For example, believing that the rustle in the grass is a dangerous predator when it is only the wind doesn't cost much, but believing that a dangerous predator is the wind may cost an animal its life.* (Shermer, Patternicity 2008)

This also may explain why American culture breeds conspiracy theories. It doesn't cost much to hear them out. If they're wrong, it is no big deal. But what if they *are* right? The cost of enduring Chicken Little is microscopic compared to the cost of ignoring Cassandra (ironically, Shermer provides the above example in an article that argues against such risk-averse thinking). If chaos theory is correct and there is a sensitive dependence upon initial conditions, then given our country's roots in conspiratorial groups, such as the Sons of Liberty, the modern American love of conspiratorial thinking makes even more sense.

Cognitive Illusions and Fallacies

The formative work of psychologists Tversky and Kahneman has inspired extensive study of the phenomena known as cognitive illusions. These cognitive illusions are epistemological anomalies. Cognitive illusions fall into three basic categories—illusions in *thinking*, in *judgment*, and in *memory*. Illusions in thinking are typically due to the misapplication of a particular heuristic or rule-of-thumb; illusions in judgment are usually due to non-logical, unconscious biases that affect the choices we make; and illusions in memory are due to problems with either the encoding or retrieval of remembered information.

For example, attributing a false cause, or believing that you've control over or can influence random outcomes, or even looking for or seeing only those pieces of evidence that confirm a theory are all thinking illusions. The illusion that guns are more dangerous than medical malpractice when, in fact, more people die of medical malpractice than of gunshot wounds is an example of a judgment illusion. When an incorrect use of a term is not noticed or when external information is remembered incorrectly because of its association with internal representations, it's a memory illusion. For example, when asked "how many animals of each kind did Moses take to the ark?" people usually answer "two" instead of noting that it was 'Noah' and not 'Moses' who took animals to the ark. Memory illusions might also explain reports of 'alien' abductions.

(For more information on cognitive illusions and the criteria to judge something as a cognitive illusion, please refer to *Cognitive Illusions: A Handbook on Fallacies and Biases in Thinking, Judgment, and Memory* (Pohl 2004). Here's a list of some of the many fallacies and illusions that riddle our reasoning, an understanding of which will help you identify them in the arguments of others and avoid them in yours: undistributed middle, illicit major and minor, composition (that which is true of parts must be true of whole), division (that which is true of whole must be true of parts), affirming consequent, denying the antecedent, false alternative, ignoratio elenchi, tu quoque, ad hominem, ad baculum, ad populum, ad verecundiam, ad misericordium, ad ignorantium, petitio principia, and plurimum interrogationum.)

Enthymemes

Another cause for faulty reasoning is unstated assumptions—conditions taken for granted to be true but not stated explicitly. In an Aristotelian syllogism, explicitly stated premises lead to a conclusion. If the explicitly stated premises are true, the conclusion will follow. There is nothing implicit except for the veracity of the premises. Arguments with implicit assumptions influencing the conclusion are called enthymemes. Aristotle categorizes an enthymeme as a rhetorical syllogism—its aim being persuasion rather than demonstration. If a premise is left unstated because it is obvious, there is no cause for concern. However, persuasion at times depends on suppressing dubious premises. Advertisements are good examples of enthymemes. The bright images and large print tell you one story. The small print or fine print that follows is the suppressed premises, which if you take the trouble to go through, will change the story, maybe, completely.

Enthymemes lead to misinterpretation because of possible ambiguities. The defense against enthymemes and the resulting model slippage is to be aware of rhetorical tools.

RHETORIC: THE THING-AS-IT-IS-COMMUNICATED

> *There grew up in Athens a body of knowledge about how to get people on your side voluntarily. This body of knowledge speedily became, and remained for more than 2,000 years, the core of Western education. It was called 'rhetoric.' (Rhetor was the usual term in Greek for 'politician.') It taught you how to get people's attention and how to argue your case once you had it.* (Lanham 2006, 25)

Rhetoric is considered the master art of the *trivium* with the mastery of both grammar and logic as prerequisites. While grammar is ultimately concerned with correctness and logic with truth, rhetoric is measured by its effectiveness. Rhetoric is the art of persuasion. Mastery of this art and its tools will not only enable you to see anomalies in the epistemic hand-me-downs you encounter but also free your own arguments of fallacies that arise from a misuse of rhetorical tools.

Anomaly: Revolutionary Knowledge in Everyday Life

Polysemes

> *Since a word is a symbol, an arbitrary sign whose meaning is imposed on it, not by nature, not by resemblance, but by convention, it is by its very nature subject to ambiguity; for, obviously, more than one meaning may be imposed on a given symbol. In a living language, the common people from time to time under changing conditions impose new meanings on the same word...* (Joseph 2002, 34)

One of the biggest sources of trouble in reasoning is ambiguities caused by the use of polysemes–terms that have many related meanings. Logical fallacies that specifically arise from ambiguity and the use of polysemes include equivocation, amphiboly, and the fallacy of accent. Equivocation is the fallacy that arises when a polyseme is used in its multiple meanings to develop an argument causing confusion. Amphiboly occurs when a sentence is constructed in a way that creates double meanings. And finally, the fallacy of accent happens when the wrong word in a sentence is emphasized creating a false impression. That master of rhetoric, the hypnotist, uses this fallacy to influence people; of course, the hypnotist doesn't call it a fallacy but an 'embedded command' (more on hypnotism in the Chapter 5).

This is why an understanding of classical rhetoric is so important, and it is here that we complete our revolution, back where we began, back to the slippage between models and reality (as if in a circuit.) This circular, or rather spiral, path is what I call the *rhetorical revolution*.

> *I hope that my speculative portrayal of analogy as the lifeblood, so to speak, of human thinking, despite being highly ambitious and perhaps somewhat overreaching, strikes a resonant chord in those who study cognition. My most optimistic vision would be that the whole field*

of cognitive science suddenly woke up to the centrality of analogy, that all sides suddenly saw eye to eye on topics that had formerly divided them most bitterly, and naturally – indeed, it goes without saying – that they lived happily ever after. (D. R. Hofstadter 2001)

Tropes or Figures of Speech

Trope is a rhetorical device that creates a shift in the meaning of a term or terms. It is meaning that we must vet carefully since the entire structure of our knowledge rests upon a foundation of analogy and metaphor. As Lakoff and Johnson say, "Metaphor is pervasive in everyday life, not just in language but in thought and action…the way we think, what we experience, and what we do every day is very much a matter of metaphor" (Lakoff and Johnson 1980, 3). The important thing to note, for purposes of mental self-defense, is that trope changes the intension of a term for the purpose of persuasion or manipulation.

Categorically, there are four master tropes–metaphor, metonym, synecdoche, and irony. Metaphor is the equation of two terms to provide meaning to one or both of them. When Pat Benatar tells us "Love is a battlefield," she is using the term 'battlefield' to describe 'love,' thereby extending the intension of the latter term to the former. Included in the master trope of metaphor are simile, onomatopoeia, personification, and antonomasia. Metonymy is the act of substituting a term or phrase for another closely related one. For example, the classic quotation "The pen is mightier than the sword" actually means that persuasion is more powerful than coercion. The pen is substituted for persuasion and the sword for coercion. Synecdoche is the substitution of a part for the whole, the whole for a part, a species for the genus, and vice versa. For example, when I ask for a Kleenex, I am not asking for a particular kind of tissue; I am simply asking for any similar sort of tissue. Similarly, the media often uses 'The White House' to mean the Executive Branch of the United States

Government. Both the concepts of branding in advertising and sound bites in the media thrive on synecdoche. Lastly, irony is simply saying one thing while meaning the opposite. In irony, there is an incongruity between the expression and the intention. For example, when you say that something was "as clear as mud," you are being ironic.

All of these tropes can leave our hard-earned knowledge and important communications vulnerable. The rhetoric of science is crucial to our understanding of the world as is the philosophy of science. As analogy and metaphor are central to all of our mental processes, the study of rhetoric will help us understand our world and ourselves better.

> *What, then, is truth? A mobile army of metaphors, metonyms, and anthropomorphisms–in short, a sum of human relations which have been enhanced, transposed, and embellished poetically and rhetorically, and which after long use seem firm, canonical, and obligatory to a people: truths are illusions about which one has forgotten that this is what they are; metaphors which are worn out and without sensuous power...* (Nietzsche 1873)

As mentioned earlier, knowing reality is an ongoing process. Here's how it works: something in reality (an anomaly, maybe) warrants our attention. In order to process this thing, a representation, symbol, model, or metaphor is established in our mind (as first-hand or 'tacit' knowledge). We reason out this model or symbol integrating it into our knowledge (for example, via the *trivium*). This knowledge is then transmitted in the form of explicit knowledge (that is, second-hand knowledge expressed using symbols and rhetoric). As a result of the representational relationship between symbols and reality, as well as the use of trope, anomalies arise and so it spirals ad infinitum.

CHAPTER 5

HYPNOSIS: RHETORIC-IN-ACTION

Hypnosis is the black sheep of the family of problems which constitute psychology. It wanders in and out of laboratories and clinics and village halls like an unwanted anomaly. (Jaynes 2000, 379)

Hypnosis is a trance-like state usually induced by a procedure comprised of a series of preliminary instructions and suggestions. It's an induced state of consciousness marked by increased suggestibility and receptivity to direction conducive to effecting behavioral change. Hypnosis effects these behavioral changes via the application of persuasive speech, that is, rhetoric.

Persuasive speech is always a function of a man's combinatorial deftness with symbols. Rhetoric, as we've seen, is the art of using language to persuade another to behave as you desire. Hypnosis simply uses rhetoric under special conditions to effect more reliable and predictable behavioral change. For example, when someone comes to the hypnotist to quit the habit of smoking, the hypnotist uses language effectively to persuade the subject to quit. Decades of addiction obliterated after just an hour or two of applied language. For hypnotism to be effective, however, the hypnotist should enable the person undergoing hypnosis to maintain strong *focus* and complete *relaxation*. (Fans of HBO television series *True Blood* may find it interesting that the etymological root of the vampire power of 'glamouring'–a variant of the Hollywood version of hypnosis–is 'grammar.')

Since 1843, when Dr. Braid coined the term 'hypnosis,' it has been well understood that the two prime drivers of hypnosis are the subject's physical *relaxation* (what he termed 'nervous sleep') and mental *focus* ('abstraction' in Dr. Braid's terms) (Braid 1853).

Nearly everyone has been in this state of hypnosis at some point or another, although we may not be aware of it. When you are home, watching an engrossing movie or your favorite TV show or when you are driving, you are likely experiencing some level of hypnosis, as you are both very relaxed and very focused. It's no coincidence then that the advertising industry capitalizes heavily upon the use of television, billboard, and drive-time radio media during these particular times. Like the hypnotist, television or billboard advertisers often use commands (which they call 'slogans'), such as 'Drink Coke' or 'Just Do It.' In this state of relaxation and focus–also referred to as 'trance,' suggestions can easily bypass our critically reasoning mind, unless of course, we are vigilant.

As Julian Jaynes says, hypnosis is a black sheep or, if you prefer, an 'anomaly.' As such, there have been several attempts at explaining and integrating this phenomenon, including the common denial that it exists. Some explain away hypnosis as merely a variant of what philosopher Jean Paul Sartre called *mauvaise foi* (aka 'Bad Faith'), a phenomenon when a person denies his or her total freedom and chooses to be like an inert entity. A core claim of existentialist thought is that humans cannot escape the freedom of choice and hypnotism is *mauvaise foi* because the person submitting to hypnosis voluntarily denies his or her total freedom, instead choosing to behave as the hypnotist suggests.

> *Sartre, as we noted earlier, says that one 'puts oneself into mauvaise foi as one puts oneself to sleep'; it is a 'spontaneous determination of our being'...One may even speculate that the analogy between putting oneself into mauvaise foi and putting oneself to sleep goes deeper. In both there is purposeful alienation of the self from concerns which might normally be those of the self.* (Fingarette 2000, 98-99)

The extent of freewill that human beings have is dubious given the work of Benjamin Libet and the Minnesota Twins Studies mentioned earlier. While Sartre is not totally off the mark, choice becomes real only if the person making a choice is aware of the options available–to go into trance or not or not make a choice at all. However, there are numerous examples of people being hypnotized without their knowledge or consent (we will come back to this shortly).

The appearance of authority is an important ingredient in effective hypnotism. A rather infamous example of how the power of authority can manipulate behavior is the Milgram Experiment devised by psychologist Stanley Milgram and conducted at Yale University in the early 1960s. To quantify how obedient people were to authority figures, clinicians (the actual unsuspecting subjects of the study) were told to administer progressively more painful and debilitating electroshocks to the apparent subject of the research. In fact, the 'victim' was actually in on the duplicity, being an actor pretending to be the victim of electroshocks. In Milgram's own words, "the point of the experiment is to see how far a person will proceed in a concrete and measurable situation in which he is ordered to inflict increasing pain on a protesting victim" (Milgram 1974, 4). He found that 65% of his clinicians willingly administered increasingly dangerous electroshocks to a 'victim' who clearly wanted the shocks to stop and who would later succumb to the shocks altogether. Simply because the supervising 'scientific' authority kept encouraging the clinicians to continue the coercive torture, they did so.

> ...*ordinary people, simply doing their jobs, and without any particular hostility on their part, can become agents in a terrible destructive process. Moreover, even when the destructive effects of their work become patently clear, and they are asked to carry out actions incompatible with fundamental standards of morality, relatively few people have the resources needed to resist authority.* (Milgram 1974)

The results of this study have been replicated several times and meta-analysis by Dr. Thomas Blass of the University of Maryland reports that this 60-65% rate of obedience is quite consistent (Blass 1999). There are other studies like Milgram's, the 'Stanford prison experiment' for example, and they all yield a similar result–our behavior can be modified by simply shifting the meanings of our models of authority.

Such seduction by the trappings of authority is the result of a conditioned response. When famous Russian psychologist Ivan Pavlov noticed that dogs salivated not just in the presence of food but also in the presence of their feeder, he devised an experiment to create associations that would yield the desired behavior. By ringing a bell every time dog food was served, the dogs (using inductive reasoning) began to confuse correlation with causation, to the point where the mere ringing of a bell, even in the absence of food, would cause the dogs to salivate. In essence, Pavlov manipulated the dogs' expectations or model of how reality worked and demonstrated he could, in this way, elicit a conditioned physiological response.

Princeton psychologist Julian Jaynes put forth an interesting theory of consciousness that is relevant at this point in the discussion of authority as it relates to hypnotism. According to Jaynes, consciousness hasn't always manifested itself as the self-reflexive phenomenon we are familiar with today. He maintained that before we made the evolutionary advance and adopted this type of consciousness, human beings possessed what he called 'bicameral consciousness.' These early people's behavior was largely the result of auditory hallucinations or the 'voices of the gods,' and hypnosis works because:

> *...it engages the general bicameral paradigm which allows a more absolute control over behavior than is possible with consciousness.* (Jaynes 2000, 379)

The hypnotist takes the place of the auditory hallucinations or authoritative 'voice of god.'

> ...*a very particular archaic authorization which also determines in part the different nature of the trance. For here, instead of the authorization being an hallucinated or possessing god, it is the operator himself. He is manifestly an authority figure to the subject. And if he is not, the subject will be less hypnotizable, or will require a much longer induction or a much greater belief in the phenomenon to begin with (a stronger cognitive imperative). Indeed, most students of the subject insist that there must be developed a special kind of trust relationship between the subject and the operator.* (Jaynes 2000, 393-394)

This belief in authority also underpins the mechanism of the *placebo effect*. When the recipient of an ineffective or even sham treatment from a doctor (an authority) *believes* that it works, the treatment brings about the expected results. fMRI studies of the brain suggest that the placebo effect and hypnosis have a lot in common (Bausell 2007). Though fMRI studies are far from perfect, Bausell has stumbled upon part of what it is that makes hypnotism work—belief (that is, paradigm).

The placebo effect, though still anomalous, is today widely recognized and accepted. The topic has been covered extensively, notably in Dylan Evans book *Placebo* (2004).

> *In a particularly striking study, patients who had undergone tooth extraction were treated with ultrasound to reduce the postoperative pain. Unknown to both doctors and patients, the experimenter had fiddled with the machine, and half the patients never received the ultrasound...the test was truly double-blind. After their jaws were massaged with the ultrasound applicator, the patients were asked to indicate their level of pain on a line where one end was labeled 'no pain' and the other 'unbearable pain'. Compared with the untreated con-*

> trol group, all those treated with the ultrasound machine reported a significant reduction in pain. Surprisingly, it didn't seem to matter whether the machine had been switched on or not. Those who had been massaged with the machine while it was turned off showed the same level of pain reduction as those who had received the proper treatment. In fact, when the ultrasound machine was turned up high, it was actually reported as giving less pain relief than when it was switched off. (Evans 2004, 28)

Note the emphasis on the term 'double blind.' Neither the patients *nor* their doctors knew about the placebo to avoid revelation through unconscious, non-verbal cues. In his book, Evans shares studies that have shown the doctor's knowledge affecting the outcome. Placebo studies reveal the power of our expectations, beliefs, and trust (that is, inductive models of reality) to shape not just our own life, but the lives of others too.

The placebo effect has also been likened to a conditioned response not unlike Pavlov's dogs salivating at the sound of bells. After repeatedly feeling better when administered medicine, the color, shape, taste, person administering the medicine, and so on, become associated with feeling better. Then these things start producing the same effect as the real medicine just like the sound of a bell did for Pavlov's dogs even in the absence of food (Evans 2004, 82). In this way, human symbols and terms too are stimuli that elicit a conditioned reflexive response. The following is a particularly compelling passage from *The Manipulation of Human Behavior*, a book based on a study sponsored by the U.S. Air Force:

> *Symbols of science can be used in a magical way, as much of the 'brainwashing' literature illustrates. Various writers have invested the techniques of interrogators with the magic of science by attaching technical labels to what actually have been traditional and pragmatic practices. In*

> *assuming the attitude of the 'hard-headed' scientist toward the problem, there is a danger in falling into an equivalent misuse of science. This would be the case were one, in effect, to attempt to counter those who present a diabolical image of the 'brainwasher' by invoking superior scientific deities to frighten this specter away. Thus, magical thinking and projections, as has been indicated, pervade prevalent judgments regarding the significance of the behavioral alterations that interrogators can effect. By substituting impassive scientific names for ordinary language with its intense connotations for human values, the impression may be given of eliminating not only these extravagant judgments but also almost all the human significance of these effects. In this way, for example, 'treachery' can become mere 'attitude change' or 'a shift in the subject's frame of reference.'* (Biderman and Zimmer 1961, 6)

Hypnotic rhetoric leverages the implicit trust given to authority. Milgram's findings clearly point to the impact of implicit belief and trust in authority.

> *Men who are in everyday life responsible and decent were seduced by the trappings of authority, by the control of their perceptions, and by the uncritical acceptance of the experimenter's definition of the situation into performing harsh acts.* (Milgram 1974, 123)

Myths

A popular myth about hypnosis is that if you are highly suggestible, then you're weak-minded. In reality, the factors important in hypnosis are the very factors that help you score high on standardized tests, the benchmark of competence and learning in our culture–focus, the ability to relax, and imagination. Suggestibility and learn-ability are actually concomitant, not mutually exclusive.

> *The really crucial point which the whole history of hypnotism demonstrates is that people most susceptible to hypnotic states are normal people. Hypnotism has never been very successful in treating the severely mentally ill.* (Sargant 1974, 31)

Another myth is that we cannot be hypnotized against our will and, therefore, anti-social hypnosis is a myth. In the 20th century, a gentleman named Milton H. Erickson, M.D., created a revolutionary paradigm shift in hypnosis and was instrumental in getting hypnosis approved by both the American and British Medical Associations. His largest contribution came in the form of his discovery of covert, conversational hypnotism that didn't require any fancy watch-swinging (known as *Chevreul's Pendulum*) or counting backwards from ten. It largely comprised of intentionally boring storytelling, with deliberate use of rhetoric intermittently, to facilitate a state of relaxation and focus and to stimulate imagination by subtly using ambiguity and the informal fallacy of accent (or 'embedded commands'). Ironically, it was Erickson, the master teacher of covert hypnotism, who popularized the myth that you cannot be hypnotized against your will or be hypnotized to commit crimes or to hurt yourself (Erickson 1939). Erickson was nearly alone in this claim as most of his contemporaries had confirmed otherwise repeatedly elsewhere.

> *...if proper rapport is established, then at least a certain number of subjects can literally be fooled into committing a crime.* (G. Estabrooks 2000, 207)

Estabrooks wasn't alone in contradicting Erickson. Loyd Rowland of the University of Tulsa, another contemporary of Erickson's and author of *Will Hypnotized Persons Try To Harm Themselves or Others?* (1939), had successfully hypnotized subjects to throw what they believed to be sulfuric acid on others, as well as grab venomous striking snakes that they didn't know were behind invisible, protective glass (For more case studies, please see Brenman (1942) and W. Wells (1941)). Dr. Paul J. Reiter, perhaps the world's leading researcher on anti-social hypnotism at the time and the author of *Antisocial or Criminal Acts of Hypnosis: A Case Study* (1958), as well as Wesley Wells, a well-respected researcher and champion of hypnotism were with Estabrooks in questioning Erickson's extraordinary claim.

> *The main point of my criticism of you* (Erickson) *is that you boldly assert that, since you have failed to get certain results, such results cannot be obtained. The fallacy of such an argument should be obvious to you now, as it was to Estabrooks.* (W. Wells 2000, 228)

While the busting of this particular myth can be disconcerting, there is a silver lining. The good news is that it is difficult to hypnotize someone against their will when they are aware that they may be hypnotized. Knowledge of this potential chink in our mental armors can serve to protect ourselves better.

In whatever way you account for the anomaly called hypnosis, whether as a variant of the placebo effect, as accessing the bicameral consciousness, as obedience to authority (this is not, by any means, an exhaustive list of explanations), or as some combination of them all, rhetoric plays a central role in achieving the intended effect.

Also, rather uncanny is the use of the term 'induction' in both hypnotic literature and epistemology. While inductive reasoning forms the basis of our beliefs (or models of reality), hypnotic induction is a pathway to changing a person's beliefs and, thus, their behavior. Both 'inductions' shape our views of the world profoundly. The defense against these unwanted intrusions is simple awareness, to be on the alert.

> *...there is a rapidly accumulating body of scientific evidence which strongly suggests that many forms of psy-*

> chotherapy and alternative medicine may be pure placebos. It is by no means difficult to access this evidence – a five minute search on the Internet will suffice to unearth several relevant documents. Psychotherapists and alternative practitioners who are unaware of the various critical studies are burying their heads in the sand. Those who are aware of them, but do not inform their patients of the fact that the treatments they are offering may be pure placebos, are infringing the principle of informed consent. (Evans 2004, 184)

We will further address the *scientism* of psychiatry in Chapter 8, but for now, just remember the best effective mental defense against manipulation is to develop keen awareness of the impact of uncritically reviewed beliefs, authority, and rhetoric.

> As the base rhetorician uses language to increase his own power, to produce converts to his own cause, and to create loyal followers of his own person, so the noble rhetorician uses language to wean men from their inclination to depend on authority, to encourage them to think and speak clearly, and to teach them to be their own masters. (Szasz, Myth of Psychotherapy 1988, 20)

CHAPTER 6

EPISTEMIC HAND-ME-DOWNS: MEMORY, RHETORIC, AND HISTORY

Any assessment of the accuracy of memory requires some record of the to-be-remembered events themselves. One way to get those records is to obtain immediate first-hand accounts of experiences that are likely to give rise to vivid recollections later. On the morning after the explosion of the space shuttle Challenger in 1986, it occurred to me that shock of hearing about this disaster might be just such an experience for many Americans. With this in mind I asked a number of Emory undergraduates to make written records of how they had heard the news on the previous day. Three years later, we compared their still vivid recollections of that experience with those records. I expect to find errors – at least minor ones...As you will see, the actual results far exceeded my expectations. (Neisser and Harsch 2000)

Less than ten percent of those surveyed by Neissar and Harsch had recollections that still matched their written records, but what's scary is the confidence each had that their memory was correct! Some even maintained that their current recollection was correct despite being shown their original written record. If knowledge from memory can degrade this much within a span of just three years, what does this indicate about the fidelity of historical 'knowledge' from centuries ago?

Even if an attempt were made by the historian to present an unbiased description of the times, fidelity to the original message degrades over time as noise creeps into the channel. You may be familiar with the Telephone Game, where players line up and the person at the beginning whispers a message to the next in line, which is then passed on from one player to the next until the last person in the line shouts out the phrase. The fun of the game is that the original message, by the time it gets to the last person, will have mutated beyond recognition. A popular story is how a World War I message from the front started out as "Send reinforcements, we're going to advance" and became "Send three and four pence, we're going to a dance" by the time it reached Headquarters.

History, whether of the oral or written variety, is subject to this same sort of distortion. It's disturbing to think about the reliability (or unreliability) of the second-hand information that we call history. How do we know whether something really happened or not? We really do not. We can only speculate about what must have happened given corroborating evidence from various streams of study, such as archeology, linguistics, genetics, and geology. If we maintain anything other than a radical agnosticism about the 'events' of history, the odds are that we are in error. The best we can do with regards to second-hand or explicit knowledge is archeology or source documents and the worst, oral history.

If the historical narrative we have is not reliable, then we lose much of the basis for our inductive reasoning. Further, most people are not too particular about their epistemology. Many people hold very strong beliefs that they truly know very little about (and conversely, there are many that perhaps discount their first-hand knowledge in deference to the prevailing paradigm). As Stanley Milgram's study demonstrated, authority can easily seduce and manipulate people by controlling their perceptions and shaping their models of reality. Charles Van Doren in his book *The History of Knowledge* (1991) describes how the vast empires of ancient times–Egypt, India, Mesopotamia, Persia, China–all had laws which governed their people but kept the rulers and law-makers beyond the scope of their own law. While people had some amount of security from other people like themselves, there was nothing that shielded them from the violence of their rulers. The common man had no choice. Their reality was to rule or be ruled. So, is history simply the surviving rhetoric of propaganda of the powerful or, at most, the story of the powerful as seen by the historian?

Adding to the already murky waters of history is the secrecy of Black Chambers. David Kahn, who was granted unprecedented access to governmental records, including those of the ultra-secret National Security Agency (NSA), describes in his book *The Codebreakers* (1967) how aristocracies developed special offices called Black Chambers to decipher intercepted coded messages from other nations. Today, the CIA, the NSA, MI6, Mossad, and the like are the Black Chambers, and they are a testament to the power of words and symbols to shape history—what would history be like if the many coded messages that were intercepted were not and vice versa? Read Thomas B. Allen's book *Declassified: 50 Top-Secret Documents That Changed History* (2008) , and you cannot help but wonder what secret epic events have not made it into the public record of history? How would knowledge of these events change our beliefs and expectations? Despite the fact that these Black Chambers perform an important function in our culture, we are nagged by Roman poet Juvenal's question 'who will police the police?'

Former President Harry S. Truman's revelations about CIA may further strengthen this question.

> *I think it was a mistake. And if I'd known what was going to happen, I never would have done it. I needed...the President needed at that time a central organization that would bring all the various intelligence reports we were getting in those days, and there must have been a dozen of them, maybe more, bring them all into one organization so that the President would get one report on what was going on in various parts of the world.*
>
> *Now that made sense, and that's why I went ahead and set up what they called the Central Intelligence Agency.*

> *But it got out of hand. The fella...the one that was in the White House after me, never paid any attention to it, and it got out of hand. Why, they've got an organization over there in Virginia now that is practically the equal of the Pentagon in many ways and I think I've told you, one Pentagon is one too many.*
>
> *Now, as nearly as I can make out, those fellows in the CIA don't just report on wars and the like, they go out and make their own, and there's nobody to keep track of what they're up to. They spend billions of dollars on stirring up trouble so they'll have something to report on. They've become...it's become a government all of its own and all secret. They don't have to account to anybody.*
>
> *That's a very dangerous thing in a democratic society, and it's got to be put a stop to. The people have got a right to know what those birds are up to. And if I was back in the White House, people would know. You see, the way a free government works, there's got be a housecleaning every now and again, and I don't care what branch of the government is involved.*
>
> *Somebody has to keep an eye on things...And when you can't do any housecleaning because everything that goes on is a damn secret, why, then we're on our way to something the Founding Fathers didn't have in mind. Secrecy and a free, democratic government don't mix.*
>
> *You have got to keep an eye on the military at all times, and it doesn't matter whether it's the birds in the Pentagon or the birds in the CIA.* (M. Miller 1973)

While history's several secrets make historical hand-me-downs suspect, common perceptions of current events and personalities are hardly true to reality. The following quote is from a recent speech by John Pilger, twice the recipient of Britain's Journalist of the Year Award:

Epistemic Hand-Me-Downs

The clever young man who recently made it to the white house is a very fine hypnotist. Partly because it is indeed extraordinary to see an African-American at the pinnacle of power in the land of slavery. However, this is the twenty-first century and race together with gender, and even class, can be very seductive tools of propaganda. For what it is so often overlooked, and what matters, I believe, above all, is the class one serves. George Bush's inner circle, from the State Department to the Supreme Court, was perhaps the most multiracial in Presidential history. It was "PC" par excellence. Think Condolezza Rice, Colin Powell. It was also the most reactionary.

Obama's very presence in the White House appears to reaffirm the moral nation. He's a marketing dream. But like Calvin Kline or Benneton, he is a brand that promises something special, something exciting, almost risque. As if he might be radical. As if he might enact change. He makes people feel good. He's a post-modern man with no politcal baggage. And all that's fake.

In his book "Dreams From My Father" Obama refers to the job he took after he graduated from Columbia in 1983. He describes his employer as, "a consulting house to multinational corporations". For some reason he doesn't say who his employer was, or what he did there. The employer was Business International Corporation which has a long history of providing cover for the CIA with covert action and infiltrating unions on the left. I know this because it was especially active in my own country Australia.

Obama doesn't say what he did at Business International and there may be absolutely nothing sinister but it seems worthy of inquiry and debate as a clue to perhaps who the man is.

During his brief period in the Senate, Obama voted to continue the wars in Iraq and Afghanistan. He voted for the Patriot Act. He refused to support a bill for single-payer health care. He supported the death penalty. As a Presidential candidate he received more corporate backing than John McCain. He promised to close Guantanamo as a priority but instead has excused torture, reinstated military commissions, kept the Bush gulag intact, and opposed habeas corpus.

Daniel Ellsberg, the great whistle-blower, was right I believe when he said that under Bush a military coup had taken place in the United States giving the Pentagon unprecedented powers. These powers have been reinforced by the presence of Robert Gates, a Bush-family crony and George W. Bush's powerful Secretary of Defense, and by all the Bush Pentagon officials and Generals who have kept their jobs under Obama.

In the middle of a recession, with millions of Americans losing their jobs and homes, Obama has increased the military budget. In Columbia, he is planning to spend $46 million on a new military base that will support a regime backed by death squads and further the tragic history of Washington's intervention in that region.

In a pseudo-event in Prague, Obama promised a world without nuclear weapons to a global audience mostly unaware that America is building new tactical nuclear weapons designed to blur the distinction between nuclear and conventional war. Like George Bush, he used the absurdity of Europe threatened by Iran to justify building a missile system aimed at Russia and China.

In another pseudo-event at the Annapolis Naval Academy, decked with flags and uniforms, Obama lied that America had gone to Iraq to bring freedom to

> that country. He announced that the troops were coming home. This was another deception. The head of the army, General George Casey, says with some authority that America will be in Iraq for up to a decade. Other generals say fifteen years.
>
> Chris Hedges, a very fine author of "Empire of Illusion" puts it very well, "President Obama" he wrote, "does one thing, and brand Obama gets you to believe another. This is the essence of successful advertising. You buy or do what the advertiser wants because of how they make you feel." And so you are kept in a perpetual state of childishness. He calls this 'junk politics.' (Pilger 2009)

As John Pilger points out, the very essence of successful advertising is to get you to believe in one thing even when the reality may be quite contrary, and advertising is not limited to products you buy but can include the packaging of public personalities and events, including wars. Later in this book, we will investigate several anomalous events in history and media reporting that have been, perhaps prematurely, swept under the rug. These unusual issues certainly challenge many commonly accepted paradigms that many of our fellow citizens are trapped within.

Historical knowledge, a case of epistemological hand-me-downs, is thus tainted not only by massive signal degradation but also by deliberate secrecy and distortion. Human history, without the benefit of strong corroborating evidence and artifacts, does not amount to more than story-telling and story-telling, is a poor foundation to base one's inductive reasoning upon.

CHAPTER 7

THOUGHT CONTROL VS. FIRST AMENDMENT

Voice or no voice, the people can always be brought to the bidding of the leaders. That is easy. All you have to do is tell them they are being attacked, and denounce the pacifists for a lack of patriotism and exposing the country to danger. (Hermann Goering, quoted in Gilbert 1947)

Words like freedom, patriotism, and security evoke strong emotional responses, revealing our paradigms–paradigms that are handed down to us, that many of us accept without question, that hide contradictions and blind us to anomalies, which can be lethal. Bertrand Russell illustrates one such contradiction:

> *The armed forces of one's own nation exist–so each nation asserts–to prevent aggression by other nations. But the armed forces of other nations exist–or so many people believe–to promote aggression. If you say anything against the armed forces of your own country, you are a traitor, wishing to see your fatherland ground under the heel of a brutal conqueror. If, on the other hand, you defend a potential enemy State for thinking armed forces necessary to its safety, you malign your own country, whose unalterable devotion to peace only perverse malice could lead you to question…*
>
> *And so it comes about that, whenever an organisation has a combatant purpose, its members are reluctant to criticise their officials and tend to acquiesce in usurpations and arbitrary exercise of power which, but for the war mentality, they would bitterly resent. It is the war*

> *mentality that gives officials and governments their opportunity. It is therefore only natural that officials and governments are prone to foster war mentality.* (Russell 1952, 51)

The First Amendment to the United States Constitution guarantees among other rights, the freedom of speech and freedom of the press. However, the history of the United States has witnessed regular infringements upon these rights. Sadly, these serious violations of the First Amendment of the Constitution were already occurring as early as the Civil War. It was during this conflict the Espionage and Sedition Acts were adopted. As Ted Galen Carpenter, Vice President for Defense and Foreign Policy Studies at the Cato Institute notes:

> *More than 2,000 people were prosecuted and 1,055 convicted under those laws – the overwhelming majority for merely criticizing the government. It is testimony to the real motive for the legislation that not a single enemy spy was ever convicted of violating the Espionage Act.* (Carpenter 1995, 26)

World War I brought even more controls upon the free flow of language, words, and symbols. This war saw established the Committee on Public Information, the Espionage Act, the Trading with the Enemy Act, and the Creel Commission. The Creel Commission's sole purpose was to influence American public opinion to support American intervention in World War I. World War II only tightened the noose around the throats of dissident voices with the advent of Roosevelt's Office of Censorship and the First War Powers Act (which authorized the complete censorship of any communication between the U.S. and any foreign country). War has often been an excuse for governments to infringe upon the freedom of speech so that the few may gain control of the many.

Another factor that limits the flow of information and thought is media monopolies. Media, in the form of television, radio, and the Internet factors enormously into how we perceive our world. In 1983, Pulitzer Prize-winning journalist Ben H. Bagdikian wrote a book called *The Media Monopoly* alerting people to the fact that just fifty corporations controlled nearly all of the news media in the United States. By 2008, just twenty-five years later, the number has come down to *six*:

1. **Time Warner, Inc.** (*AOL, New Line Cinema, Time, Warner Bros, CNN, HBO, Turner Broadcasting, The CW Network, People Magazine, Sports Illustrated, DC Comics, The WB Network, Warner Brothers, The Cartoon Network*)

2. **Walt Disney** (*ABC, Touchstone, The Disney Channel, Pixar, Miramax, ESPN*)

3. **Viacom** (*CBS, MTV, Nickelodeon, Showtime, Simon & Schuster, Paramount Pictures, Blockbuster, DreamWorks, Spike TV, BET, Comedy Central, Infinity Outdoor*)

4. **Vivendi Universal – NBC Universal** (*NBC, Interscope Records, Geffen Records, Def Jam Records, Motown, The Oxygen Network, The Weather Channel, The SyFy Network, Vivendi Publishing*)

5. **News Corp** (*Fox, Sky TV, 20th Century Fox, MySpace, TV Guide, HarperCollins Publisher, My Network TV, The New York Post, Dow Jones & Company, The Wall Street Journal, Barron's*)

6. **Bertelsmann AG** (*BMG music publishing, Random House, Doubleday. By the way, Bertelsmann was the single largest producer of Nazi propaganda during WW II*)

When you compare the timeline of these consolidations with the decline of the perceived reliability of the news, the results are revealing. In 1985, 55% Americans considered news stories accurate. That number has today fallen to 29%. The percentage of Americans who said that news organizations do not get their facts straight has gone up from 34% to 63%. (Pew Research Center 2009). 60% of those who participated in the survey believe news organizations to be politically biased and only 20% believe them to be free of the influence of powerful people and organizations. You can check out http://people-press.org/report/543/ for more on the decline of trustworthiness and increase in partisan biases in the American press today.

The Internet is still mostly a decentralized information resource; however, the giants like Google, Yahoo, and Microsoft are performing mergers and acquisitions like crazy, just like traditional media did. In other words, you may eventually receive only their perspective—what they have filtered for you as 'newsworthy.'

So at one end, a great majority of our second-hand knowledge and symbology is becoming more and more centralized and homogenized through corporate media consolidations, and at another end, we have the Black Chambers trying to restrict the scope of our inquiries and communication in the name of national security. We are offered a false alternative between news-bite trope or politicized Newspeak. All the while, grammar, logic, and rhetoric, those great tools of reason that help us sort through the symbolic muck, have largely been eliminated from the public education system. As our paradigms are based not only on our first-hand experience but also on the epistemic-hand-me downs of explicit knowledge, any violation of the right to freedom of speech effectively translates into active thought control.

Law: The Rhetoric of Peaceful Liberty

The root of the English very word 'war' is *werra* (of Frankish-German origin) and means confusion, while peaceful, mutual agreement is often described as 'coming to terms.' We have two basic choices with regards to how we relate to our fellow human beings–violent confusion or peaceful clarity.

> *There grew up in Athens a body of knowledge about how to get people on your side voluntarily. This body of knowledge speedily became, and remained for more than 2,000 years, the core of Western education. It was called 'rhetoric.' (Rhetor was the usual term in Greek for 'politician.') It taught you how to get people's attention and how to argue your case once you had it.* (Lanham 2006, 25)

Language can create war; it can maintain peace. The use of precise, discrete language, free of shifts in meaning not only keeps confusion away but also helps two or more parties to 'come to terms.' The alternative to a culture of peaceful coexistence is one of physically violent coercion.

One of the great themes in the history of humankind has been the tyranny of a few masters over the masses of slaves. Jefferson's concept that all men are equal before the law had to be fought for to be adopted and must still be guarded from potential oligarchs that would seek to return to the old way, by any means necessary. As Bertrand Russell writes, democracy by itself is not sufficient to prevent horrors from being perpetrated. There must be respect for the individual, the kind that inspired the doctrine of Rights of Man (Russell 1952, 64).

> *To exercise power costs effort and demands courage. That is why so many fail to assert rights to which they are perfectly entitled – because a right is a kind of power but they are too lazy or too cowardly to exercise it. The virtues which cloak these faults are called patience and forbearance.* (Nietzsche 1986, 371)

A culture that doesn't abuse its symbols with trope or leverage asymmetrical information excessively receives that wonderful and precious social commodity–*trust*.

But what happens if we are conditioned to not even trust ourselves?

CHAPTER 8

THE HELPLESSNESS PARADIGM

He who cannot obey himself will be commanded. That is the nature of living creatures. (Nietzsche 1961, 137)

Authority and power love to foster helplessness; it makes the act of domination much easier. At the turn of the twentieth century, the Rockefellers were the most powerful family in the world, and they began to practice philanthropy to promote their own particular worldview. In 1902, the Rockefeller Institute for Medical Research was created, and Dr. Simon Flexner was appointed its first director. Thanks to their association with the Rockefellers, the Flexner family would become incredibly influential in their day. The eldest brother Simon we've already met, but there was also a younger brother Bernard who would later become a founding member of the international policy powerhouse known as *The Council on Foreign Relations*. In 1910, Abraham Flexner, another younger brother, published a report that was to shape the direction of the healthcare system in the US.

Abraham Flexner, an educator (not a medical practitioner), surveyed 155 different medical schools all across North America and published the Flexner Report aimed at creating a uniform standard for medical pedagogy across the United States and Canada. The Report was co-sponsored by the Rockefeller Foundation and the Carnegie Foundation, and it issued a number of recommendations that medical schools should adopt. For example, all medical graduates had to pass exams from a state licensing board.

At the time of the initial survey, only 35 of the 155 medical schools surveyed met or surpassed Flexner's idea of what a medical institution should be. With the immense economic and political capital yielded by Rockefeller and Carnegie Foundations, it was tantamount to economic suicide to ignore the Flexner Report mandates; without conforming to the standards issued by the report, medical schools stood little chance of obtaining funding from the Rockefeller/Carnegie financial octopus. Further, thanks to the heavy political lobbying by the Carnegie Foundation and the American Medical Association, state and local licensing requirements mandated adherence to the Flexner Report's standards. (The AMA is still a heavy hitter in Political Action Committees, coming in at #14 on the all time largest contributors list since they've been keeping track (Center for Responsive Politics 2009).)

It has been argued that the Flexner Report raised the standards of medical education and practice and established medicine on a firm scientific foundation correcting many of the deficiencies present then (Mindrum 2006). However, government involvement through licensure led to the cartelization of medicine, and America is now facing the consequences. Medical training and care have become costly, and the supply of doctors does not meet demand, raising not only the income of doctors but also their status, sometimes to god-like proportions. This creates a fertile ground for the informal reproduction of Milgram's 'obedience to authority' experiment on innocent victims, this time called 'patients,' frequently with tragic results.

According to the Journal of the American Medical Association (the most widely circulated peer-reviewed medical journal on the planet), iatrogenesis is the third largest killer of Americans after heart disease and cancer (Starfield 2000, 483). Iatrogenesis is "[a]ny injury or illness that occurs as a result of medical care" (Taber's Medical Dictionary). That's right, medical malpractice or 'death by doctor' is *the third leading cause of death in the United States.* The author of the article, Dr. Barbara Starfield from the Johns Hopkins School of Hygiene and Public Health, reveals the following annual mortality statistics:

The Helplessness Paradigm

Deaths	Cause
106,000	Non-error, negative effects of drugs
80,000	Infections in hospitals
45,000	Other errors in hospitals
12,000	Unnecessary surgery
7,000	Medication errors in hospitals
225,000	**Total Deaths by Doctor per Year**

Death by doctor kills nearly a quarter million Americans each year. Starfield's statistics were computed conservatively compared to an independent report published by the National Academy of Science's Institution of Medicine. Per the IOM report, every single year, deaths due to iatrogenic causes range from 230,000 to 284,000 (Starfield 2000, 484). To put that in perspective, iatrogenic deaths account for five times the deaths caused by motor vehicles,[1] almost seventeen times the deaths caused by firearms,[2] and over 75 times the number that died in the 9/11 attack on the Twin Towers.

A comparison of the United States healthcare against that of 13 other countries revealed its abysmal conditions. According to Starfield, the United States scored:

Rank	Health Indicator
13 (last)	Low birth-weight percentages
13 (last)	Neonatal mortality and infant mortality overall
11	Post neonatal mortality
13 (last)	Years of potential life lost (excluding external causes)
11, 10, 10, 7, 3	Life expectancy for females at 1, 15, 40, 65, 80 years respectively
12, 12, 9, 7, 3	Life expectance for males at 1, 15, 40, 65, 80 years respectively
10	Age-adjusted mortality

---

[1] 43,664 according to http://www.cdc.gov/nchs/FASTATS/acc-inj.htm.
[2] 12,791 according to http://www.cdc.gov/nchs/fastats/homicide.htm.

Japan, Sweden, and Canada take the gold, silver, and bronze medals, respectively, for having the best healthcare systems today. Only Germany ranks worse than the United States for the quality of its healthcare. It's interesting to note here that the two men who were the most influential in the development of the American healthcare system, Simon and Abraham Flexner, basically modeled their vision for America after the German medical system considered to be the best at that time. However, after a century of application, both systems have yielded the worst and second-worst ranking of all. The overall effect of the Flexner Report hasn't been that great after all. The Rockefeller-Carnegie funded Flexner Report has brought with it a higher cost of healthcare, higher mortality rates for patients, and an incentive for the medicalization of everything. Even the supposed increase in quality is dubious.

The main redress victims have against medical malpractice and iatrogenesis is medical malpractice lawsuits. Currently, there is a large push to reduce the power of these lawsuits as part of the current healthcare reform package before Congress. It's being called 'Tort Reform,' whose advocates claim that too many of such cases are without merit (Leonhardt 2009). If it goes through, this tort reform will remove one of the largest deterrents to medical malpractice.

Dr. David L. Edsall was the Dean of the Harvard Medical School and was trained under the new homogenizing protocols instituted by the Flexner Report. In *The New Drug Story*, Edsall describes his experiences:

> *I was, for a period, a professor of therapeutics and pharmacology, and I knew from experience that students were obliged then by me and by others to learn about an interminable number of drugs, many of which were valueless, many of them useless, some probably, even harmful...Almost all subjects must be taken at exactly the same time, and in almost exactly in the same way by all students, and the amount introduced into each course is such*

> *that few students have time or energy to explore any subject in a spirit of independent interest. A little comparison shows that there is less intellectual freedom in the medical course than in almost any other form of professional education in this country.* (Bealle 1958)

Apart from the cartelization of medicine and its disastrous consequences (to the unsuspecting public, of course), the imposition of Flexner Report recommendations affect today the autonomy of medical providers.

> *If a third party joins a two-person game, its rules ipso facto are altered. If an insurance company or the state becomes an active participant in the business of defining what counts as medical care as well as in paying for and distributing the service, the nature of medical practice undergoes a process of metamorphosis: the economics of medicine, the power relations between doctor and patient, and the very meaning of all the key terms change. This is familiar territory. 'He who pays the piper calls the tune'. Troy Duster, professor of sociology at the University of California, Berkeley, phrases it more professionally: 'Once the third party steps in to pay the physician for delivering health-care to the patient, the interests of the third party will typically supercede those of the patient.'* (Szasz 2001, 39)

In this context, a short note on the German medical system is in order. Flexner's model was based on the German model as we have already mentioned. Otto Von Bismarck, the first Chancellor of the unified German Empire and practically the inventor of the modern welfare state as we know it, implemented legislation that mandated health insurance, as well as accident, disability, and retirement insurance in Germany at the end of the nineteenth century. Rudolf

Virchow, the Father of Pathology and the most prominent physician in Germany at the time, saw the politics behind the Sickness Insurance Law (1884) and opposed Bismarck. (Virchow founded the disciplines of cellular and comparative pathology, which along with other modalities, such as medical imaging and blood testing, are the bedrocks of scientific medical diagnosis used today.)

> *Despite his passionate interest in preventive medicine, Rudolf Virchow, the foremost medical man of the day, as well as a Reichstag deputy, opposed the measure. He did so because he realized that its aim was political, not medical. Bismarck hated communism. He introduced socialized medicine into Germany to buy the loyalty of the German masses: Bismark adopted 'nationalistic socialism to end international socialism.'...[He] was the first leader of a great nation to fight Communism by adopting Communism...[His scheme] became an important feature of the German militaristic state; it helped pave the way for Hitler a generation later.' I would add that Bismarck's nationalization of medicine paved the way not only for Hitler but also, more specifically, for the Nazi program of medicalized mass murder, and, more generally, for the triumph of the therapeutic state and the prevailing political incorrectness of opposing it.* (Szasz 2001, 67)

The policies inspired by the Rockefeller-Carnegie funded Flexner report has led us to a system that produces an uncomfortably large number of cases of iatrogenesis. This 'open secret' has been discussed candidly in medical literature, as you can see here:

> *The Flexner Report, published in the early 20th century, turned medicine toward a firm scientific foundation and raised standards of education and practice. This corrected many of the profession's deficiencies present at the turn of*

> the century such that medicine became capable of improving the health of humanity. While the focus of education on the sciences suited the needs of the era, the pendulum may have swung too far. As medical schools clamored for funding from wealthy capitalists to achieve new standards, they lost autonomy and adopted unsaid values that were possibly a danger to humanity. This dynamic may have led to the manifestation of medicine's dark history marked by the eugenics movement and the Tuskegee Project. This history demonstrates how medicine can impact humanity detrimentally when the broader scope of the humanities and arts is lost. In spite of this understanding, medical education has been dogmatic and resistant to change. The paper ends with a description of the crisis in modern medicine that is on par with the problems it faced in the early 1900s and concludes that it is indeed time for another revolution. (Mindrum 2006)

While one may argue that iatrogenesis is an unintended consequence of Rockefeller's Flexner Report, it is hard to deny the very intentional funding of the American Eugenics Society by the Rockefellers, the Carnegies and the Harrimans. Eugenics attempts to eliminate genetic defects and improve the qualities of a species or race by controlling reproduction so that only individuals with supposedly desirable traits reproduce. *In the Name of Eugenics* by Daniel J. Kelves and *The War against the Weak* by Edwin Black go into meticulous detail regarding the players in the eugenics movement and their atrocities, including forced segregation, sterilization, and mass-murder. In fact, Mrs. E.H. Harriman (wife of the wealthy railroad magnate), John D. Rockefeller (oil magnate), and Andrew Carnegie (steel magnate) funded nearly the entire Eugenics movement in the United States (Kelves 1995, 208).

> *When we were done, we had assembled a mountain of documentation that clearly chronicled a century of eu-*

> *genic crusading by America's finest universities, most reputable scientists, most trusted professional and charitable organizations, and most revered corporate foundations. They had collaborated with the Department of Agriculture and numerous state agencies in an attempt to breed a new race of Nordic humans, applying the same principles used to breed cattle and corn. The names defined power and prestige in America: the Carnegie Institution, the Rockefeller Foundation, the Harriman railroad fortune, Harvard University, Princeton University, Yale University, Stanford University, the American Medical Association, Margaret Sanger, Oliver Wendell Holmes, Robert Yerkes, Woodrow Wilson, the American Museum of Natural History, the American Genetic Association and a sweeping array of government agencies from the obscure Virginia Bureau of Vital Statistics to the U.S. State Department.* (E. Black 2003, xxii)

While those that are prone to Type I errors might begin to make a connection between eugenicist philosophy, Rockefeller funding, and the incredible amount of iatrogenic deaths, those folks that tend to make Type II errors might not be able to say anything since they might have been among the 225,000 that die annually due to medical malpractice.

> *Only after the truth about Nazi exterminations became known did the American eugenics movement fade. American eugenic institutions rushed to change their names from eugenics to genetics. With its new identity, the remnant eugenics movement reinvented itself and helped establish the modern, enlightened human genetic revolution. Although the rhetoric and organizational names had changed, the laws and mindsets were left in place. So for decades after Nuremberg labeled eugenic methods genocide and crimes against humanity,*

The Helplessness Paradigm

> *America continued to forcibly sterilize and prohibit eugenically undesirable marriages.* (E. Black 2003, xvii)

Now, you have a twofold incentive for the medicalization (a form of scientism) that has been institutionalized by the medical industry. Medicalization not only generates unearthly amounts of revenue, but kills those that some may see as 'prone to weakness.' In other words, it's a wealth-seeking eugenicist's (like Rockefeller, Carnegie, Harriman) dream! The Flexner Report has created a brilliant revenue generating iatrogenic model where the victim not only asks for 'healthcare,' but demands it!

> *In Nazi Germany, the entire medical profession was corrupted, systematically to justify the euthanasia program. In the United States today, the entire medical profession is corrupted, systematically lying to justify a variety of medical interventions—especially for sexual and 'mental' problems and the need for pain relief.* (Szasz 2001, 47)

Before we move on to the largest medical trope being perpetrated on the American public, 'psychiatry,' here are some facts to ponder regarding the overlapping Venn diagrams of medicalization, iatrogenesis, and the eugenics program.

- The very first Noble Prize winner to come out of the Rockefeller Institute, Alex Carrel, was a brilliant surgeon, a Nazi sympathizer, and an intractable eugenicist. Carrel's suggestion to use poison gas to murder predated Nazi implementation by only a few years: *A euthanasia establishment, equipped with a suitable gas, would allow the humanitarian and economic disposal of those who have killed, committed armed robbery, kidnapped children, robbed the poor or seriously betrayed public confidence... Would the same system not be appropriate for lunatics who have committed criminal acts?* (Carrel 1935)

- The patent holder to Zyklon-B–the poisonous gas used to murder the prisoners at Auschwitz-Birkenau in Nazi Germany–was I.G. Farber, the fourth largest corporation in the world at the time. I.G. Farber was founded in 1925 through a merger of six other companies, one of which was the German pharmaceutical giant Bayer. In 2006, Bayer was outed for selling a vaccine in the 1980s that was *accidentally* contaminated with HIV that led to hemophiliacs contracting HIV due to the vaccine.

- Paul Hoch was a Hungarian psychiatrist trained in Germany. He became the Senior Clinical Psychiatrist at the Rockefeller-supported New York State Institute for Psychiatry at Columbia in 1933. In the 50s, he led a research project for the CIA on LSD. Later, Governor Nelson Rockefeller appointed Hoch to the position of Commissioner of Mental Hygiene for the state of New York, and that brings us to psychiatry, an entire branch of medicine based upon trope.

Psychiatric Trope

Of all tyrannies, a tyranny exercised for the good of its victims may be the most oppressive. It may be better to live under robber barons than under omnipotent moral busybodies. The robber baron's cruelty may sometimes sleep, his cupidity may at some point be satiated; but those who torment us for our own good will torment us without end for they do so with the approval of their own conscience. They may be more likely to go to Heaven yet at the same time likelier to make a Hell of earth. Their very kindness stings with intolerable insult. To be 'cured' against one's will and cured of states which we may not regard as disease is to be put on a level of those who have not yet reached the age of reason or those who never will; to be classed with infants, imbeciles, and domestic animals.

The Helplessness Paradigm

(Lewis 1970; cited in Rothbard and Hoppe 2003)

Central to the proliferation of the field of psychiatry in the United States was the Rockefeller Foundation again. Dr. David L Edsall, Dean of the Harvard Medical School and a Rockefeller Foundation trustee, made a report which was "at once both the charter and the point of departure for this new venture" (Fosdick 1989). According to Dr. Edsall, psychiatry was *"the most backward, the most needed, and potentially the most fruitful field in medicine."* (Fosdick 1989) It would indeed be economically fruitful for the medical industry if 'mental cases' could be brought under the auspices of medical doctors. So it is that as eugenics began to fade out of favor, especially as the German atrocities up to and during World War II became better known, the Rockefeller Foundation too began to shift its focus from genetic 'hygiene' toward 'mental' hygiene. The Rockefeller Foundation, along with other foundations, has put in many millions of dollars into the medicalization of normal mental processes.

> *Except for a few objectively identifiable brain diseases, such as Alzheimer's disease, there are neither biological or chemical tests not biopsy or necropsy findings for verifying or falsifying DSM* (The Diagnostic and Statistical Manual of Mental Disorders) *diagnoses.* (Szasz 2008, 2)

The Diagnostic and Statistical Manual of Mental Disorders (commonly known as the DSM and which is publishing volume V soon) is published by The American Psychiatric Association and is the 'bible' of mental health care professionals. There are no statistics though, only 'diagnoses'– labels, terms, analogies. The DSM is a mostly unscientific lexicon of diagnoses where mental 'diseases' are voted in, or voted out. For example, due to political pressure the anti-woman diagnosis of 'hysteria' was voted out, and in 1973, the mental 'disease' of homosexuality was determined, by vote, to no longer be a 'disease.' If only we could simply vote out cancer or AIDs from being a disease! At their core, the DSM and APA are not scientific; they are political. Have you ever wondered why an oncologist (a cancer doctor) would prescribe chemotherapy to cure himself of cancer, but why you'll never find a psychiatrist that has prescribed for himself a lobotomy or electo-shock therapy? Psychiatrists have a suicide rate considerably higher than that of the general public, yet you'll seldom find a psychiatrist taking SSRI to treat 'depression.' Why?

> *Let us be honest: Classifying nondiseases as disease serves the economic, existential, and professional interest of the classifiers and is, to boot, socially expected of them. For the vast majority of health-care professionals–especially the legions of psychologists, social workers, grief counselors, drug abuse specialists, and other practitioners of existential cannibalism–it would be professional suicide to categorize nondiseases correctly: their every instinct of self-interest opposes such truth telling. After incessantly inflating the concept of illness, medical professionals and the media have, in effect, lost their ability to call a spade a spade and a heart a nonspade.* (Szasz 2001, 37)

Psychiatry's revenue-generating ability is illustrated by the near doubling of the rate of anti-depressant prescriptions in the United States between 1995 and 2005. According to Mark Olfson and Steven Marcus, fifteen million more Americans were on anti-depressants than just a decade before (Olfson and Marcus 2009). The rise in the usage of these pharmaceuticals is such that in Britain drinking water is now tainted with them. As environmental spokesman Norman Baker puts it, "This looks like a case of hidden mass medication of the unsuspecting public and is potentially a very worrying health issue" (Dawar 2004).

And this massive increase has taken place despite research that shows these anti-depressants offer no more help than placebo, only dangerous side-effects.

> *Meta-analyses of antidepressant medications have reported only modest benefits over placebo treatment, and when unpublished trial data are included, the benefit falls below accepted criteria for clinical significance.* (Kirsch, et al. 2008)

This means there was effectively zero difference in positive performance between the drug and a placebo in patients with moderate depression and a **clinically insignificant** difference among patients with severe depression. So all these patients would have had the same benefits had they taken a sugar pill without exposing themselves to dangerous side-effects and without the expense of the medications. The website Drugs.com lists the potential side-effects of SSRI anti-depressants, such as Prozac:

> *Get emergency medical help if you have any of these signs of an allergic reaction: skin rash or hives; difficulty breathing; swelling of your face, lips, tongue, or throat.*
>
> *Call your doctor at once if you have any new or worsening symptoms such as: mood or behavior changes, anxiety, panic attacks, trouble sleeping, or if you feel impulsive, irritable, agitated, hostile, aggressive, restless, hyperactive (mentally or physically), more depressed, or have thoughts about suicide or hurting yourself.*
>
> *Call your doctor at once if you experience any of these serious side effects from using Prozac:*
>
> - *severe blistering, peeling, and red skin rash;*
> - *very stiff (rigid) muscles, high fever, sweating, fast or uneven heartbeats, tremors, overactive reflexes;*
> - *nausea, vomiting, diarrhea, loss of appetite, feeling unsteady, loss of coordination; or*

- *headache, trouble concentrating, memory problems, weakness, confusion, hallucinations, fainting, seizure, shallow breathing or breathing that stops...*

(For a complete list of side effects, please visit http://www.drugs.com/prozac.html.)
...or you can take a sugar-pill, and incur none of these risks. To further drive home the point, the International Coalition of Drug Awareness has compiled over 3,200 documented violent crimes committed (mostly after the year 2000) by individuals taking SSRI anti-depressants, like Prozac. They report over 700 murders, 200 completed murder-suicides and other acts of violence, 62 road rage tragedies, 48 school shootings/incidents, 47 cases of postpartum depression, and more (for more information, visit: http://www.ssristories.com/index.html).

> *I am alarmed at the monster that Johns Hopkins neuroscientist Solomon Snyder and I created when we discovered the simple binding assay for drug receptors 25 years ago. Prozac and other antidepressant serotonin-receptor-active compounds may also cause cardiovascular problems in some susceptible people after long-term use, which has become common practice despite the lack of safety studies.*
>
> *The public is being misinformed about the precision of these selective serotonin-uptake inhibitors when the medical profession oversimplifies their action in the brain and ignores the body as if it exists merely to carry the head around! In short, these molecules of emotion regulate every aspect of our physiology.* (Pert 1997, 8)

The fraud being perpetrated by psychiatry is one of the biggest 'open' secrets of our culture today. The most fearless and brilliant critic of psychiatry and its tropes has been the psychiatrist

The Helplessness Paradigm

Thomas Szasz. In fact, his criticisms and the strength of his arguments have become central to the very subject of 'mental illness' (see http://plato.stanford.edu/entries/mental-illness). Fortunately, Szasz is not alone. Psychiatrists Alvin Pam (Assistant Professor of Psychiatry at Albert Einstein College of Medicine), Colin A. Ross (Director of the Dissociative Disorders Unit at Charter Hospital of Dallas), and others are holding the feet of psychiatry to the fire, challenging their peers to come to terms that psychiatry is not a science.

In a 1972 study, Stanford psychologist David Rosenhan had perfectly healthy volunteers admit themselves to a mental hospital where the hospital psychiatrist then interpreted their normal behavior to be insane. Rosenhan writes about his experiment, "It is clear that we cannot distinguish the sane from the insane *in* psychiatric hospitals" (1973). One year later, Donald Naftulin, a young psychiatrist, hired an actor to impersonate a psychiatrist before 53 subjects (composed of psychiatrists, psychologists, social workers, and educators). Journalist Mark Oppenheimer summarizes the findings nicely:

> *For 'The Doctor Fox Lecture: A Paradigm of Educational Seduction,' a 1973 article still widely cited by critics of student evaluations, Donald Naftulin, a psychiatrist, and his co-authors asked an actor to give a lecture titled "Mathematical Game Theory as Applied to Physician Education." The actor was a splendid speaker, his talk filled with witticisms and charming asides – but also with "irrelevant, conflicting and meaningless content." Taking questions afterward, the silver-haired actor playing "Dr. Myron L. Fox" affably answered questions using "double talk, non sequiturs, neologisms and contradictory statements." The talk was given three times: twice to audiences of psychiatrists, psychologists and social workers, the last time to graduate students in educational philosophy.* (Oppenheimer 2008)

Sadly, the knowledge of the fraud of psychiatry is not new, just suppressed. As early as 1879, Nellie Bly, an investigative journalist wrote an article for the New York Times entitled "Tormenting the Insane" (1879) that detail her findings when she faked 'mental illness' to gain admittance to an insane asylum under an assumed name to write an expose. Famed German imposter Gert Postel even faked being a psychiatrist and got away with it.

> *As far as psychiatry is concerned, it can be said that if you're able to perform linguistic acrobatics you can make a career for yourself. That is what Psychiatry is based on.*
> (Postel 2001)

Both iatrogenesis and psychiatry feed off of asymmetrical information and conditioned responses developed after a lifetime of exposure to suggestions via authority figures like doctors, intellectuals, and military and law enforcement. Only awareness can prevent the damage they cause, but the responsibility of spreading that awareness depends largely upon word of mouth (which can be unreliable due to signal degradation), as there are no 'Rockefellers,' 'Carnegies,' or 'Harrimans' funding public service announcements to educate people about the dangers of iatrogenesis or psychiatry, with the noted exception of the Citizen's Commission on Human Rights (http://www.cchrint.org).

Given the depth to which psychiatric trope has permeated our culture, it would do us well to understand psychiatry's political aims, which leaders in the profession had made clearly early on. John Rawling Rees, co-founder of the World Federation for Mental Health and the Tavistock Institute, says:

> *Public life, politics and industry should all of them be within [psychiatry's] sphere of influence…we have made a useful attack upon a number of professions. The two easiest of them naturally are the teaching profession and the Church; the two most difficult are law and medicine.*
> (Rees 1940)

The Helplessness Paradigm

THE INSTITUTIONALIZATION OF LEARNED HELPLESSNESS

> *To achieve world government, it is necessary to remove from the mind of men their individualism, loyalty to family traditions, national patriotism, and religious dogmas.* (Chisholm 1790)

'Learned Helplessness,' a term coined by Martin Seligman (a psychologist working with dogs in respondent conditioning), is a specific kind of depression where, after a series of unpredictable horrible events, the subject begins to feel an overwhelming sense of helplessness and the inability to come up with methods to escape such events, even if escape is possible. This explains how a 5-ton elephant can be confined by nothing more than a wispy chain–it believes it cannot escape because when young, it repeatedly tried to and failed, and because it believes it cannot escape, it doesn't even try.

> *Martin Seligman developed the concept of inoculation from stress from his famous studies of learning in dogs. He put dogs in a cage that had an electric shock pass through the floor at random intervals. Initially the dogs would jump, yelp, and scratch pitifully in their attempts to escape the shocks, but after a time they would fall into a depressed, hopeless state of apathy and inactivity that Seligman termed 'learned helplessness.' After falling into a state of learned helplessness, the dogs would not avoid the shocks even when provided with an obvious escape route.* (Grossman 1995, 81)

In 1992 and 1993, the Harwood Group conducted a study of students for the Kettering Foundation to help explain how young people felt about politics and their role in the process. The findings of the 'Kettering Study,' as it became known, could be summed up with the simple helpless phrase: "Why bother voting, it won't do any good....." The study found that these students thought they could participate in the political process only in three general ways (voting, the signing of petitions, and protesting), none of which offered any escape from the status quo in their view. This *helplessness* is akin to Nietzsche's concept of 'slave morality'–where the values of weakness and subservience are extolled–or of Nils Bejerot's 'Stockholm Syndrome'–a victim's perverted sense of loyalty to their victimizer.

If the work of B.F. Skinner (the father of operant conditioning), Ivan Pavlov, Jacques Loeb (another German trained M.D. and Rockefeller faculty member), and other behaviorists has taught us anything, it is that human beings can be conditioned and trained to learn anything, from killing to helplessness. Modern militaries exploit Pavlovian and Skinnerian conditioning to overcome natural human aversion to murder (Grossman 1995).

> *It is to be expected that advances in physiology and psychology will give governments much more control over individual mentality than they now have even in totalitarian countries. Fichte laid it down that education should aim at destroying free will, so that, after pupils have left school, they shall be incapable, throughout the rest of their lives, of thinking or acting otherwise than as their schoolmasters would have wished...Diet, injections, and injunctions will combine, from a very early age, to produce the sort of character and the sort of beliefs that the authorities consider desirable, and any serious criticism of the powers that be will become psychologically impossible. Even if all are miserable, all will believe themselves happy, because the government will tell them that they are so.* (Russell 1952, 61-62)

The Helplessness Paradigm

The Pedagogy of Surrender

> *In our dreams, people yield themselves with perfect docility to our molding hands. The present educational conventions [intellectual and character education] fade from our minds, and unhampered by tradition we work our own good will upon a grateful and responsive folk. We shall not try to make these people or any of their children into philosophers or men of learning or men of science. We have not to raise up from among them authors, educators, poets or men of letters. We shall not search for embryo great artists, painters, musicians, nor lawyers, doctors, preachers, politicians, statesmen, of whom we have ample supply. The task we set before ourselves is very simple ...we will organize children ...and teach them to do in a perfect way the things their fathers and mothers are doing in an imperfect way.* (Gatto 2000)

This excerpt from the first mission statement of Rockefeller's General Education Board reveals the Rockefeller agenda. Education is one of the most effective modern methods of propaganda–simply control knowledge to maintain power. The American power-elite understood this all too well. As the 1991 New York State Teacher of the Year and three-time winner of the New York City Teacher of the Year award John Taylor Gatto writes:

> *By 1917, the major administrative jobs in American schooling were under control of a group referred to in the press of that day as 'the Education Trust.' The first meeting of this trust included representatives of Rockefeller, Carnegie, Harvard, Stanford, the University of Chicago, and the National Education Association. The chief end, wrote the British evolutionist Benjamin Kidd in 1918, was to 'impose on the young the ideal of subordination.'*

> *At first, the primary target was the tradition of independent livelihoods in America. Unless Yankee entrepreneurialism could be put to death, at least among the common population, the immense capital investments that mass production industry required for equipment weren't conceivably justifiable. Students were to learn to think of themselves as employees competing for the favor of management. Not as Franklin or Edison had once regarded themselves, as self-determined, free agents.* (Gatto 2000, 274)

This is not an indictment of the Harriman, Rockefeller, or Carnegie families as their philanthropy has indeed yielded much good. This is more of a warning about the centralization of influence, of symbology, and the power of trope. It's not yet hopeless. Martin Seligman's studies show how 'learned helplessness' can be evaded–solely by an awareness of the possibility of escape. Even a single instance of escape inoculated Seligman's dogs against learned helplessness.

> *Other dogs were given a means of escape after receiving some shocks but before falling into learned helplessness. These dogs learned that they could and would eventually escape from the shocks, and after only one such escape they became inoculated against learned helplessness. Even after long periods of random, inescapable shocks these inoculated dogs would escape when finally provided with a means to do so.* (Grossman 1995, 81)

There is No 'Conscience' Without 'Science'

What kind of economy did rhetorical education imply? A market economy, obviously enough. An economy that depends on persuasion. It is the rhetorical habit of mind that creates both the free market and the free market of ideas. The freedom comes from persuasion, not co-

ercion, whether you buy a product or an idea. Planned economies constrain attention; rhetorical markets attract it. They do not compel agreement; they invite it. (Lanham 2006)

Adam Smith, largely remembered as the father of economics, was appointed the chair of logic and *rhetoric* at Glasgow University in 1751 and the chair of moral philosophy in 1752. So Smith was very well versed in both rhetoric and moral philosophy by the time he published his famous 1776 book *A Brief Inquiry into the Nature and Causes of the Wealth of Nations* that launched the study of economics. Before that however, in 1759, he wrote "The Theory of Moral Sentiments" which argued that morality and justice hinged not upon the pronouncements of religious authorities, but upon sympathy. For Smith, the very act of observing others makes people self-aware of the consequences of their own actions. The Stanford Encyclopedia of Philosophy explains Smith's theory of sympathy thus:

> *As spectator of an agent's suffering we form in our imagination a copy of such 'impression of our own senses' as we have experienced when we have been in a situation of the kind the agent is in: "we enter as it were into his body, and become in some measure the same person with the agent.*

In this way, we view others as an analog to ourselves. Adam Smith's theory of moral sentiments has come to be called by psychologists today the Theory of Mind. Simply put, the Theory of Mind is the ability to infer, without any real demonstrable proof, another person's perspective (since we can't effectively prove that others have minds). The Theory of Mind is the natural assumption that your own mind is analogous to the minds of others through the observation of the emotions, actions, and language of others. Ethics, in addition to epistemology, is a function of analogy.

> *The connection between ethics and the scientific understanding of consciousness, while rarely made, is ineluctable, for other creatures become the objects of our ethical concern only insofar as we attribute consciousness (or perhaps potential consciousness) to them. That most of us feel no ethical obligations toward rocks—to treat them with kindness, to make sure they do not suffer unduly—can be derived from the fact that most of us do not believe that there is anything that it is like to be a rock.*
> (Harris 2005, 174)

This is where we find the basic meaning behind the Golden Rule (or Kant's categorical imperative, if you prefer): *Do unto others as you would have them do unto you.* As a side note, the originator and propagators of the Golden Rule meme clearly didn't account for masochism. So, in light of the existence of masochists, the Golden Rule should be reformulated as follows:

Treat others as they would like to be treated unless:

1. You don't have any clue how they would like to be treated, or
2. Treating them how they wanted to be treated would jeopardize yourself or innocent others. In those cases, treat them how you would like to be treated.

As I reiterate throughout this book, I raise these unusual points to simply raise awareness of the power of paradigms and anomaly, even in everyday life. Of course, not all doctors, psychiatrists, psychologists, politicians, high-ranked military personnel, media executives, and philanthropists are bad; in fact, I'd believe that most of them have good intentions. That said, the excuse of 'unintended consequences' has been overly used, even abused. I do believe that there are some in positions of power who *do have bad intentions* (as determined by the reformulated Golden Rule above), and it would be naïve to say otherwise. Only an awareness of the problems (that is, anomalies) associated with our analogies or worldviews (that is, paradigms) can lead to resolution (that is, revolution). As I. Bernard Cohen writes in *Science and The Founding Fathers*:

> *In fact, one of the traditional uses of analogies in the physical and biological sciences, and in the earth sciences, has been to serve as a tool of discovery. Jeremy Bentham declared analogy is a major tool for scientific discovery. In the Origin of Species, Charles Darwin not only made extensive use of analogies but also indicated that although he was aware that 'analogy may be a deceitful guide,' he was able to conclude 'from analogy that probably all the organic beings which have ever lived on this earth have descended from some one primordial form, into which life was first breathed.*
>
> *This kind of use of analogy illustrates what Alfred North Whitehead has called the 'logic of discovery,' the path to increasing knowledge, which Maxwell called the process of 'exciting appropriate mathematical ideas.' We may take note that in the advancement of all the sciences, in the way that one scientist makes use of the creation of an important new concept in the form of what seems to be a partial or imperfect replication of the original or even a dramatic recasting of the original...*
>
> *In introducing the concept of 'logic of discovery,' Whitehead contrasted it with what he called the 'logic of the discovered.' In this case, an analogy is used once a discovery has been made. There are three primary uses of analogy under such circumstances. One is to help explain a difficult concept or principle... Thus, analogy can be used to make something understandable or understood.*

A somewhat similar use of analogies is to validate a novel concept, to make it seem plausible... Thus, analogy can be used to support something, to make it seem reasonable, to make it acceptable and accepted. (Psychiatry, couched in scientism, is a good example.)

Closely related to this use of analogy is one in which the likeness is presented as carrying with it a system of values. ***Because the utilization of analogy moves it further into the realm of rhetoric and because in this function the comparison may involve a single word or image, that is, a brief figure of speech, as well as an extended representation of similarities, analogy employed in this third way may perhaps be designated as a metaphor, even though a clear and strict distinction between analogy and metaphor is neither necessary nor useful.*** *What is especially significant is the transfer of value systems...*

Examples from within the fields of the strictly mathematical and physical and natural sciences may come to mind less readily because during the Scientific Revolution of the seventeenth century rhetoric fell into disfavor among thinkers and writers whose prose style was influenced by the 'new science' or the 'new philosophy,' as it was sometimes called. The advocates and practitioners of the 'new philosophy' held that rhetoric had no place in scientific discourse. Science was to be presented in simple language, with clear descriptions of the evidence of experiment and observation, followed by strict induction or deduction. Each step was to be set forth in language that was unadorned and clearly understood—without rhetorical flourishes to distract the reader from the evidence and the logic. This was one of the reasons for the great esteem given to mathematics, which seemed to be the most rhetoric-free form of discourse. Today, however, a num-

ber of historians and philosophers of science have come to recognize that the sciences have a rhetoric their own, a set of conventions and assumptions and even rules of discourse which are not provable by experiment or observation or guaranteed by mathematics. (Parenthesis and emphasis added) (Cohen 1997, 27-30)

Part II

New Word Order – Paradigms Of Trope & Centralization

CHAPTER 9

SECRET PARADIGMS

Conspiracy is as natural as breathing. And since the struggles for advantage nearly always have a rhetorical strain, we believe that the systematic contemplation of them forces itself on the student of rhetoric. Indeed, of all the motives in Machiavelli, is not the most usable for us his attempt to transcend the disorders of his times, not by either total acquiescence or total avoidance, but by seeking to scrutinize them as accurately and calmly as he could? (Burke 1962, 166)

An old joke goes, "just because you are paranoid doesn't mean they aren't out to get you." So given all the tricks and lessons of the mind that we've previously covered, how do we make sense of things? How do we ensure that we haven't somehow unwittingly donned a proverbial tin-foil hat and fallen prey to some wacky conspiracy theory in error? In other words, how do we avoid making up scenarios in our minds when there aren't any there? How do we avoid Type I errors or false positives of this nature?

Recent research by Jennifer Whitson and Adam Galinsky has shown, through a series of experiments, that individuals who feel that they lacked control over their lives were more likely to make Type I errors–seeing images that weren't there, developing superstitions, and perceiving conspiracies where none existed. It seems that we have a built-in bias to simply create answers, regardless of their veracity, to questions when we don't know the real answers. This seems congruent with the research of Harvard University biologist Kevin R. Foster and University of Helsinki biologist Hanna Kokko mentioned earlier–it's better to suspect a dangerous predator than wind when the grass rustles. Explicit knowledge is anyways unreliable, so when it comes to anomalous 'conspiracies,' suspecting a predator is better than losing your life.

However, on the other hand, the Pentagon Papers are a perfect example of legitimate conspiracy. Daniel Ellsberg, a young, bright Harvard economics PhD and former Marine, served under Secretary of Defense Robert McNamara at the Pentagon (the subject of the stellar documentary *The Fog of War* by Errol Morris) during the Vietnam War. Later, he became one of the hyper-influential think-tank the RAND Corporation's experts on the Vietnam conflict and was invited to contribute to a classified document commissioned by McNamara, which would later become known as the Pentagon Papers.

Disgusted by what he read, Ellsberg used his high-level security clearance to leak this information to the *only industry protected by the Constitution of the United States of America, the free press.* The Pentagon Papers revealed, among other things, that government decision makers knew that the war was unwinnable but would persist in the conflict despite this knowledge. Additionally, it was acknowledged that this persistence would cause many more casualties, many of which would never be made public.

The Nixon Administration's attempts to suppress this revolutionary information included burglarizing the offices of Lewis Fielding (Ellsberg's psychiatrist) to dig up dirt to discredit Ellsberg in the media. However, fortunately for the American public, they failed to find any such files. The integration of the anomalous information in the Pentagon Papers eventually led to the resignation of President Nixon.

Even more scandalous information regarding the Vietnam conflict would surface with the publication of an article with the unusual title "Skunks, Bogies, Silent Hounds, and the Flying Fish" by Robert J. Hanyok. Originally published in 2001 in the National Security Agency's classified journal *Cryptologic Quarterly*, the article was made public in 2003 under the Freedom of Information Act. Hanyok's article reveals that the Congressional resolution to authorize broad military action in Vietnam at President Lyndon B. Johnson's request was based on falsified records that created the appearance that North Vietnam had attacked American destroyers on August 4, 1964, in the Gulf of Tonkin. A similar manipulation of intelligence was used to justify the Iraq War. In 2004, National Security Archive research fellow John Prados edited a National Security Archive briefing book which published for the first time some of the key intercepts from the Gulf of Tonkin crisis.

'Unintended consequences,' a term readily accepted by the academia, the media, and politicians alike is, in many ways, simply the other side of the coin of conspiracy. The heavy casualties sustained in an unwinnable war were the 'unintended consequence' of actions based on lies, read conspiracy. As the Pentagon Papers proved, words (symbols) are powerful. And it is important to differentiate clearly between 'intentional' and 'unintentional' consequences since most modern legal systems differentiate between *mens rea* (intent) and *actus rea* (action). If you can demonstrate that there was no intent to break the law, then it was an accident and, therefore, unpunishable making 'unintended consequences' the escape hatch for perpetrators of legitimate conspiracies.

Obviously, there are plenty of kooky conspiracy theories out there; however, when confronted with actual evidence of fraud and deception (no matter how anomalous), any and all synonyms of the adjective 'kooky' should be tossed into the waste bin. The play of symbols can have serious consequences, especially when symbols become centralized, without the freedom to falsify or to present alternative paradigms. The dictator is the embodiment of this type of centralization of symbology and the very antithesis of liberty. In language, there is power; this is why the word 'dictator' has its roots in the Latin word *dicto* which means: I dictate, I repeat, I compose, or I prescribe.

CHAPTER 10

THE INFIDEL REVOLUTION

"I believe in Plexiglas," Hall said. (Associated Press, CNN 2008)

After a firefight, when his commander asked him whether he believed in God, Jeremy Hall, a 23-year-old Iraq war veteran, said "no," that he believed in Plexiglas. The Humvee on which Jeremy was a gunner had taken several bullets in its protective shield. Hall is an anomaly for those who claim that there aren't any atheists in fox holes.

If you happened to be born in India, would you practice the Hindu religion? What if you were born in Tibet or Iraq? Radio talk-show host Adam Corolla has an interesting theory of religion. He explains religious views with a sports team analogy. The sports teams you follow are largely determined by the place you are born and the teams your parents and/or friends follow; and so it is, it seems, with religion. It's amazing that the arbitrary nature of what religion you are born into doesn't convince more people to take a less confident stand with regards to religion. It's even more amazing that agnosticism is a rather controversial worldview to this day.

Because of the rarity of public discussions that politely challenge religious belief, it is important to open the discussion about this anomalous world view. Representative Fortney Hillman 'Pete' Stark, Jr., in January 2007, openly acknowledged his atheism becoming the first American congressman to do so. Soldier Jeremy Hall sued the army for harassment for being an atheist. Is the religion analogy so deeply rooted that we feel threatened by even the mention of anomalies?

It wasn't really until the reign of the George W. Bush administration that it was drummed into me exactly how anomalous the naturalistic or humanistic worldview truly was. Post-9/11 military conflicts and conversation seemed to be framed within a paradigm of religious values–'Christian freedom' versus 'Islamofascists.' Over and over again, I heard media pundits evangelizing that ours was a nation founded on Christian values. When I'd try to speak sensibly to people about this, it was as if I were speaking into an abyss. When the abyss would speak back, it would often try to convince me that God is in the gaps–that God is everything that science cannot explain–unaware that they are equating God with *anomaly*. Instead of examining these anomalies seriously, the conceptual bucket 'God' and 'faith' is used to justify this epistemic laziness. It seems that it isn't so much that 'God is dead' in the Nietzschean sense (1961); it's just that 'God' is a weak metaphor, a short cut to critical thinking.

> *When we talk about the beliefs to which people consciously subscribe–"The house is infested with termites," "Tofu is not a dessert," "Muhammad ascended to heaven on a winged horse"–we are talking about beliefs that are communicated, and acquired, linguistically. Believing a given proposition is a matter of believing that it faithfully represents some state of the world, and this fact yields some immediate insights into the standards by which our beliefs should function. In particular, it reveals why we cannot help but value evidence and demand that propositions about the world logically cohere. These constraints apply equally to matters of religion.* (Harris 2005, 51)

If you've ever debated the existence of God, you'd have found that the discussion can go on and on and the inability to falsify the existence of God is taken as proof that God exists. Bertrand Russell devised a nice analogy, referred to as Russell's Teapot, to handle claims like this:

> *If I were to suggest that between the Earth and Mars there is a china teapot revolving about the sun in an elliptical*

> *orbit, nobody would be able to disprove my assertion provided I were careful to add that the teapot is too small to be revealed even by our most powerful telescopes. But if I were to go on to say that, since my assertion cannot be disproved, it is an intolerable presumption on the part of human reason to doubt it, I should rightly be thought to be talking nonsense. If, however, the existence of such a teapot were affirmed in ancient books, taught as the sacred truth every Sunday, and instilled into the minds of children at school, hesitation to believe in its existence would become a mark of eccentricity and entitle the doubter to the attentions of the psychiatrist in an enlightened age or of the Inquisitor in an earlier time.*
> (Russell 1996)

But just as the inability to falsify the existence of God doesn't prove God exists, so it is that the absence of proof isn't proof of absence–God may exist. As with history and other second-hand knowledge, the only tenable position is the agnostic one. I admit, as religious rhetoric began to saturate the culture post 9/11, the works of Richard Dawkins and comedians like George Carlin, Bill Hicks, and Bill Maher came as welcome relief. It was a delight watching Christopher Hitchens on television dismantling the arguments of the less rhetorically gifted with regards to religion. Unfortunately, there are plenty of 'atheists' that deep down just selfishly and childishly enjoy hurting people and aren't interested in sincerely persuading theists that their position is problematic. These bullies are fairly easy to spot. They are the ones calling people names, making snide remarks, etc. all over the Internet and elsewhere. This sort of behavior only makes the non-religious look bad. A recent Gallup Panel survey (March 24-27, 2008) shows that only scientologists are disliked more than atheists (Jones 2008). If you tell a theist that their faith is wrong, you've already annoyed them. Add to that a bit of condescension or ridicule and you've lost them entirely.

Anomaly: Revolutionary Knowledge in Everyday Life

Ok, now for some religious paradigm busting anomalies.

Religions Often Call For, and Get, Murder

This is easy to verify. Think of the death threats and fatwa associated with Salman Rushdie's publication of *The Satanic Verses* or the Danish newspaper *Jyllands-Posten* Muhammad controversy where cartoonists' lives were threatened for publishing drawings of Muhammad. In Nigeria, Muslims killed Christians with machetes and burned down churches citing this Dutch cartoon. Think of the Salem Witch Trials, the Spanish Inquisition where countless numbers were murdered in the name of religion, or the myriad reports of suicide bombers. It isn't just racial pseudoscience (for example, eugenics), but religious beliefs that often bring justifications for genocide that divorce us from sympathy, making it easy to kill other human beings. In *Why I Am Not A Christian*, Bertrand Russell writes:

> *The Spaniards in Mexico and Peru used to baptize Indian infants and then immediately dash their brains out: by this means they secured these infants went to Heaven.* (1957)

When confronted with these atrocities done in the name of religion, defenders of religiosity often claim that the perpetrators were misinterpreting scripture or manipulating it to suit their ends. While this might be true, how should one interpret these words?

> *Anyone who is captured will be run through with a sword. Their little children will be dashed to death right before their eyes. Their homes will be sacked and their wives raped by the attacking hordes. For I will stir up the Medes against Babylon, and no amount of silver or gold will buy them off. The attacking armies will shoot down the young people with arrows. They will have no*

mercy on helpless babies and will show no compassion for the children. (Isaiah 13:15-18)

They are backstabbers, haters of God, insolent, proud, and boastful. They are forever inventing new ways of sinning and are disobedient to their parents. They refuse to understand, break their promises, and are heartless and unforgiving. They are fully aware of God's death penalty for those who do these things, yet they go right ahead and do them anyway. And, worse yet, they encourage others to do them, too. (Romans 1:24-32)

If a man lies with a male as with a woman, both of them shall be put to death for their abominable deed; they have forfeited their lives. (Leviticus 20:13)

Whoever strikes his father or mother shall be put to death. (Exodus 21:15)

This is certainly not good will and love. Quoted and interpreted literally, there is no manipulation. And this is just a few from a very, very long list of atrocities.

Religion is Not Always About God; It's About Power

The Family is an 'invisible' association with very public members–senators, congressmen, generals, foreign dictators, and businessmen, including arms manufacturers. Jeff Sharlet, journalist and author, expert on religious subcultures, lived undercover with brothers of The Family and reported his astonishing experiences. It all began in 1935 when an immigrant preacher named Vereide organized a small group of businessmen sympathetic to European fascism. Soon Vereide's small group had gone international and began recruiting the powerful–the new chosen.

"The Cedars (the Family headquarters) *has a heart for the poor," they like to say. By "poor" they mean not*

> the thousands of literal poor living barely a mile away but rather the poor in spirit, for theirs is the kingdom: the senators, generals, and prime ministers who coast to the end of Twenty-fourth Street in Arlington in black limousines and town cars and hulking S.U.V.'s to meet one another, to meet Jesus, to pay homage to the god of *The Cedars*. (Sharlet 2003)

In the name of Jesus, they cite Hitler, Lenin, and Mao. Doug Coe, the leader of the Family says, "We work with power where we can, build new power where we can't."

> "So," Doug told us, "my friend said to the president (of Uganda), 'Why don't you come and pray with me in America? I have a good group of friends–senators, congressmen–who I like to pray with, and they'd like to pray with you.' And that president came to The Cedars, and he met Jesus. And his name is Yoweri Museveni, and he is now the president of all the presidents in Africa. And he is a good friend of the Family." (Sharlet 2003)

In the process of meeting with Jesus, the Family effects 'quiet diplomacy' (in George H W Bush's words), sometimes benign, like leading Rwanda and Congo to a peace accord, but more often not so benign.

> During the 1960s, the Family forged relationships between the U.S. government and some of the most anti-Communist (and dictatorial) elements within Africa's postcolonial leadership. The Brazilian dictator General Costa e Silva, with Family support, was overseeing regular fellowship groups for Latin American leaders, while, in Indonesia, General Suharto (whose tally of several hundred thousand "Communists" killed marks him as

> *one of the century's most murderous dictators) was presiding over a group of fifty Indonesian legislators. During the Reagan Administration, the Family helped build friendships between the U.S. government and men such as Salvadoran general Carlos Eugenios Vides Casanova, convicted by a Florida jury of the torture of thousands, and Honduran general Gustavo Alvarez Martinez, himself an evangelical minister, who was linked to both the CIA and death squads before his own demise.* (Sharlet 2003)

Sharlet's discovery gives us an anomalous look at the lives of those who run the world. The Family has played key roles in the New Deal, the Cold War, and the 'no-holds-barred economics of globalization.' The core of the Family believe God's old covenant broken and themselves the 'new chosen.'

> *...they forge "relationships" beyond the din of vox populi (the Family's leaders consider democracy a manifestation of ungodly pride) and "throw away religion" in favor of the truths of the Family.* (Sharlet 2003)

The Founding Fathers Were Not Christians

> *People like you are not holding up the Constitution and are going against what the founding fathers, who were Christians, wanted for America!*

In a meeting of atheists and freethinkers at Camp Speicher in Iraq, Specialist Jeremy Hall was excited to see Major Freddy J. Welborn. However, Major Welborn soon began to rebuke those gathered accusing them of going against the constitution and the founding fathers' wishes. (Banerjee 2008)

That America's founding fathers were Christians upholding Christian values is another of those paradigms that people accept without question. However, facts speak differently.

- The Constitution of the United States of America, a philosophical continuation of the Magna Carta of 1215, marked the end of the paradigm of the 'Divine Right of Kings' in the Western world. The King was supposed to be anointed by God and any resistance to the King was resistance of God and therefore a sin. The Divine Right of Kings was the concept by which the monarchs of Europe justified their power over their subjects, and this concept was derived from the various instances in the Bible where God and prophets anoint kings for Israel. The Magna Carta, the Revolution of 1688 in England, otherwise known as the Glorious Revolution, and the Declaration of Independence were all events that increasingly removed 'Christian' values from the governance of men.

- A ubiquitous Christian value that the Founders sought to abolish was the institution of slavery. Again, we find justification of slavery aplenty in the Bible. Here are some excerpts:

    > *You may buy male and female slaves from among the nations that are round you. You may also buy from among the strangers who sojourn with you and their families that are with you, who have been born in your land; and they may be your property. You may bequeath them to your sons after you, to inherit as a possession forever; you may make slaves of them, but over your brethren the people of Israel you shall not rule, over one another, with harshness.* (Leviticus 2:44-46)
    >
    > *Slaves, be obedient to those who are your earthly masters, with fear and trembling, in singleness of heart, as to Christ...* (Ephesians 6:5)

    Given the endorsements of slavery in the Bible, it is important to note that, while many of the Founders owned slaves, they sought to provide the keys to their escape in these words of the original draft of the Declaration of Independence:

> *He* (King George III of Great Britain) *has waged cruel war against human nature itself, violating its most sacred rights of life and liberty in the persons of distant people who never offended him, captivating and carrying them into slavery in another hemisphere, or to incur miserable death in their transportation thither. This piratical warfare, the opprobrium of infidel powers, is the warfare of the Christian king of Great Britain.*

Though Georgia and South Carolina ensured the removal of this portion from the final draft, thus preventing the Constitutional Convention from abolishing slavery, slavery was outlawed in a majority of states within the lifetimes of the Founding Fathers, which is an incredible feat given the context of the times.

- Just in case there is any further confusion regarding the views of the Founding Fathers on 'Christian values,' here are their recorded perspectives:

   > *Millions of innocent men, women and children, since the introduction of Christianity, have been burnt, tortured, fined and imprisoned; yet we have not advanced one inch towards uniformity.* (Jefferson, Notes on Virginia 1782)
   >
   > *Christianity neither is, nor ever was a part of the common law.* (Jefferson, Letter to Dr. Thomas Cooper 1814)
   >
   > *I almost shudder at the thought of alluding to the most fatal example of the abuses of grief which*

> *the history of mankind has preserved– the Cross. Consider what calamities that engine of grief has produced!* (Adams, Letter to Thomas Jefferson 1816)

> *What influence, in fact, have ecclesiastical establishments had on society? In some instances, they have been seen to erect a spiritual tyranny on the ruins of the civil authority; on many instances, they have been seen upholding the thrones of political tyranny; in no instance have they been the guardians of the liberties of the people. Rulers who wish to subvert the public liberty may have found an established clergy convenient auxiliaries. A just government, instituted to secure and perpetuate it, needs them not.* (Madison, Memorial and Remonstrance against Religious Assessments 1785)

- If excerpts from letters and notes are not persuasive enough, here's another excerpt–this time from an official document–the 1796 US treaty with Tripoli, "Treaty of Peace and Friendship between the United States of America and the Bey and Subjects of Tripoli of Barbary," written under the presidency of George Washington and signed under the presidency of John Adams. It is historical evidence of the official denial of a Christian basis whatsoever for the U.S. Government:

> *As the Government of the United States of America is not, in any sense, founded on the Christian religion; as it has in itself no character of enmity against the laws, religion, or tranquility, of Mussulmen (Muslim); and, as the said States never entered into any war, or act of hostility against any Mahometan (Mohammedan) nation, it is declared*

> by the parties, that no pretext arising from religious opinions, shall ever produce an interruption of the harmony existing between the two countries. (Treaty of Tripoli, Article 11 1796)

The views of the Father of the American Revolution and master of revolutionary rhetoric Thomas Paine are not very different:

> Of all the systems of religion that ever were invented, there is no more derogatory to the Almighty, more unedifying to man, more repugnant to reason, and more contradictory to itself than this thing called Christianity. (Paine 1975)

The Founders steered as far as possible from Christianity, and for good reason. After the atrocities of the Church dominated the dark ages, the Founders sought better metaphors than those provided by Christianity, and they found such metaphors and analogies in the science of their time.

> *The Founding Fathers were very much aware that they were invoking metaphors and analogies from medicine and the physical and biological sciences in their political discourse.* (Cohen 1997)

Thus begins I Bernard Cohen's outstanding book *Science and the Founding Fathers,* and it ends thus:

> *Many thinkers of the Age of Reason, including some framers of the Constitution and other American political leaders or theorists had a deep conviction that there is a parallelism between the world of nature and the world of human existence. They believed, accordingly, that sound systems of government and of the organization of society should display some analogy, some set of similarities in*

> *both values and actual forms, with the systems of nature. All of the framers of the Constitution would have agreed that no system of government or of society could be sound and stable if it contravened any of the fundamental principles of nature revealed by science.* (Cohen 1997, 280)

It was perhaps the power of this scientific metaphor that led Benjamin Rush, one of the Founding Fathers, to create another political institution laded with 'scientism,' the institution of psychiatry. The difference between the scientific metaphors of the Founders and the 'scientism' of psychiatry is that the Founders did not intend to confuse politics with science, whereas psychiatry uses 'scientism' to make politics seem scientific. So with 'scientism' as it is with science; it can be used for good or for evil. We must always remember sympathy for our fellow human beings, and the Founding Fathers sought to institutionalize this sympathy.

The Infidel Revolution

THE NAZIS, HOWEVER, WERE CHRISTIAN

And while we're at it, let's shatter the belief that Nazis were atheists. Equating atheists with Hitler and the Nazis is a rather popular trope for generating hate against atheists. However, the truth is that the Nazis and proponents of their Third Reich were Christians. For Hitler, the First Reich was the Holy Roman Catholic Empire or the dark ages and the Second Reich was the welfare state of Otto Von Bismarch. Following are some documented statements that reveal the Christian faith of Hitler and the Nazis:

Hitler in a speech given on April 22, 1922: *My feeling as a Christian points me to my Lord and Savior as a fighter. It points me to the man who once in loneliness, surrounded by a few followers, recognized these Jews for what they were and summoned men to fight against them and who, God's truth! Was greatest not as a sufferer but as a fighter. In boundless love as a Christian and as a man I read through the passage which tells us how the Lord at last rose in His might and seized the scourge to drive out of the Temple the brood of vipers and adders. How terrific was His fight for the world against the Jewish poison.* (Hitler 1942, 19)

Joseph Goebbels (Reich Minister of Public Enlightenment and Propaganda for Hitler's Nazi Party) through his fictional character Michael: *Christ is the genius of love, as such the most diametrical opposite of Judaism, which is the incarnation of hate. The Jew is a non-race among the races of the earth.... Christ is the first great enemy of the Jews....that is why Judaism had to get rid of him. For he was shaking the very foundations of its future international power. The Jew is the lie personified. When he crucified Christ, he crucified everlasting truth for the first time in history.* (Steigmann-Gall 2003)

Joseph Goebbels, Evangelisches Zentralarchiv in Berlin, March 2, 1934: *When today a clique accuses us of having anti-Christian opinions, I believe that the first Christian, Christ himself, would discover more of his teaching in our actions than in this theological hairsplitting.* (Steigmann-Gall 2003)

Joseph Goebbels, in a broadcast on April 19, 1936: *We have a feeling that Germany has been transformed into a great house of God, including all classes, professions and creeds, where the Führer as our mediator stood before the throne of the Almighty.*

Heinrich Himmler (Head of the Schutzstaffel (SS), the personal guard unit for Hitler), Bundesarchiv Berlin-Zehlendorf, 28 June 1937: *In ideological training I forbid every attack against Christ as a person, since such attacks or insults that Christ was a Jew are unworthy of us and certainly untrue historically.* (Steigmann-Gall 2003, 131)

Hermann Göring (Second in Command of the Third Reich): *Although he himself [Hitler] was a Catholic, he wished the Protestant Church to have a stronger position in Germany, since Germany was two-thirds Protestant.* (International Military Tribunal 1945)

Before winding up this myth, here's an anomaly for Christians to contemplate: The Catholic Church, as history shows, was quick to excommunicate and admonish academics and scientists who presented anomalous views that challenged the Church orthodoxy, like Charles Darwin, Rene Descartes, Immanuel Kant, and John Locke, accusing them of heresy. However, atrocities of any nature and degree do not seem to warrant action from the Church. At least, to this day, no German Catholic Nazi, including Adolf Hitler, was ever excommunicated as a result of their war crimes during the Holocaust. Maybe, because it was all done in the name of Christ.

Nietzsche, Hitler's Adopted Philosopher, Was VERY Anti-Nazi

Hitler adopted Friedrich Nietzsche and his philosophy but saw only what he wanted to see. He'd visit Nietzsche's museum in Weimar often and pose for photographs of himself staring in rapture at the bust of the great man (Shirer 1981).

When I first read Nietzsche's philosophy in high school, I was drawn to his attacks upon authority, slave morality, and religious superstition and his concept of the *übermencsh* (translates into 'superman' (Nietzsche 1961)). However, I was put off by what seemed like an argument for a plurality of 'truths.' However, going back to Nietzsche later, I understood his perspectivism–all ideas originate from particular points of view (paradigms) and the value of an idea or its truth can be judged only by taking into account the perspective. While earlier this seemed to allow for plurality of truths, later I saw it as a safeguard–a warning against philosophic system-building, a warning against rigidly adopting dogmatic paradigms, like Nazism.

Not only was Nietzsche a champion of truth and reality but also of self-criticism. Further, Nietzsche's writings were purposely unsystematic and self-contradictory so as to urge his readers to think for themselves, *to discover their own paradigms.*

> *Now I go alone, my disciples, You, too, go now, alone. Thus I want it...Go away from me and resist Zarathustra! And even better: be ashamed of him! Perhaps he deceived you...One repays a teacher badly if one always remains nothing but a pupil...Now I bid you lose me and find yourselves; and only when you have all denied me will I return to you.* (Nietzsche 1961)

Opposed to blind obedience to systems of faith, especially those with a 'leader' (of a religion, philosophy, or any human system for that matter), Nietzsche doesn't sound like a supporter of Hitler. Finding one's own way as a realist and rational being cannot go hand-in-hand with a totalitarian regime.

Unfortunately, many Nazi propagandists tried to co-opt some of Nietzsche's concepts, like the 'will to power' and the *'übermencsh'* to justify their atrocities. Below are excerpts from the writings of scholars of philosophy that throw light on Nietzsche's views:

> *His [Nietzsche's] specific statements against German nationalism, moreover, are numerous and emphatic. '[The Germans] have on their conscience all that is with us today-this most anti-cultural sickness and unreason there is, nationalism, this national neurosis with which Europe is sick, this perpetuation of European particularism, of pett politics'. And then, with uncanny prescience: 'You [German intelligentsia] think that you seek truth? You seek a 'Führer' and would be glad to follow orders.'* (Lang 2002, 54)

> *Of course it didn't take long for me to realize how poorly Nietzsche's thought did in fact fit into the National Socialist world view. More than his vitriolic words against the Germans, which might, as Alfred Bäumler showed,*

> be attributed to Nietzsche's disappointment at his reception in the Fatherland, it was his massive attacks on anti-Semites that proved too much for the Nazis to swallow. In 1942, a friendly book-dealer steered me to a text just published by the Weimar Duncker press with the title "Nietzsche, Juden, Antijuden". It was by the Wagnerian Curt von Westernhagen, who, to put it briefly, presented the philosopher as a friend of the Jews and his spirit as Jewish. (Müller-Lauter 2002, 69)

> ...Nietzsche recommended (in "Human, All Too Human") the expulsion of the anti-Semites from Europe and in "Daybreak" spoke of the need to fuse the Jewish race with the other races of Europe for the advancement and betterment of European culture. (Brinker 2002, 108)

Nietzsche has been blamed for providing a philosophical paradigm that allowed Nazism and its atrocities. However, Nietzsche's writings warned against political apathy which enables 'the cold monster'–the state–and public opinion to control individual destinies (Nietzsche 1961).

> Writing in the context of the emergence of Bismarck's German Reich, Nietzsche is severely critical of 'politics' (by which he means Machtpolitik) as a way of addressing, or solving, the problem of human existence. From his early to his last writings Nietzsche's thought is characterized by an opposition between 'Geist' (spirit) and 'Reich'. What humanity needs is not a violent political revolution, but changes in education and in its ways of thinking. It needs to ground 'spirit' in a conception of 'culture'.

> In many respects Nietzsche's critique of modern politics has much in common with the political thinking of Alexis de Tocqueville (1805-59) and John Stuart Mill (1806-

73). *Like Tocqueville, for example, Nietzsche sees hidden dangers in the new political realities opened up by the modern industrial world, modern democracy, and a money-economy. Modernity for both is characterized by social atomism, moral malaise, and the cultivation of private experience and private taste at the expense of public action. This creates a political culture that is lacking in vigour. The danger of this degeneration of politics, in which politics is dominated by the class interests of the modern money-economy and by the instrumental rationality of modern technology, is that is can lead to a situation in which people lose political control over their own destinies and become politically apathetic. At this point the 'state' – the 'cold monster', as Nietzsche like to refer to it – begins to dominate political life and to cultivate the tyranny of the majority ('public opinion') at the expense of individual liberty and genuine public action (this menace, also clearly seen by Mill, is what Tocqueville referred to as 'soft despotism').*

Like Tocqueville, Nietzsche gave a pejorative flavour to liberal individualism. Both saw modern individualism as resulting in a self-centred preoccupation with purely personal ends. For Nietszche, the danger is that society will lose sight of the importance of culture and allow philistinism to take over. Society becomes made up of a herd of 'last men and women' who are concerned only with 'happiness' (understood in the sense of the satisfaction of material desires) and who cannot conceive of anything higher or nobler beyond (über) themselves. These people no longer wish to cultivate themselves, to engage in risks and experiments, but seek only a dull and safe 'bourgeois' existence. (Ansell-Pearson 1994)

In one of his last works, *Ecce Homo* (a critical assessment of his own

Birth of Tragedy), Nietzsche makes his view of life crystal clear:

> *This ultimate, most joyous, most wantonly extravagant Yes to life represents not only the highest insight but also the deepest, that which is most strictly confirmed and borne out by truth and science. Nothing in existence may be subtracted, nothing is indispensable–those aspects of existence which Christians and other nihilists repudiate are actually on an infinitely higher level in the order of rank among values than that which the instinct of decadence could approve and call good. To comprehend this requires courage and, as a condition of that, an excess of strength: for precisely as far as courage may venture forward, precisely according to that measure of strength one approaches the truth. Knowledge, saying Yes to reality, is just as necessary for the strong as cowardice and the flight from reality–as the 'ideal' is for the weak, who are inspired by weakness.* (Nietzsche 1997)

CHAPTER 11

POLITICS AS PRO-WRESTLING

The whole aim of practical politics is to keep the populace alarmed by menacing it with an endless series of hobgoblins, all of them imaginary. (Mencken 1956, 29)

The champion, a very powerful man, walks down the ramp to the deafening sound of thousands of fans cheering his name, holding up signs with his name and catch-phrases. The challenger, follows in a like manner albeit with a smattering of boos thrown in, but with his own followers and signs held up. Upon reaching the platform, both men ready themselves for battle, watching the referee for the signal to begin. Some of the fans know the 'fix' is in, while others think it's real. There is a promoter behind these men who in essence owns them and who runs the show and benefits from it. This narrative could well be describing a United States presidential election debate instead of a professional wrestling event.

Until Vincent Kennedy McMahon exposed pro-wrestling as theater, *not* competition, in the mid-eighties, professional wrestling was literally a conspiracy. Most people suspected it was fake, or at the very least 'fishy,' but very few were on the 'inside' and knew. Again, these words describe what many feel about American politics. Today, given all the vote fraud, the influence of special interest groups and political action committees, and the trope created by pollsters like Frank Luntz, the idea that American politics operates similarly to modern WWE-style professional wrestling is not that far-fetched. When ancient Roman poet Juvenal commented on how the Roman public had given up its birthright of political involvement in exchange for 'bread and circuses' provided by politicians, little did he know that the circus provided would be politics itself.

Anomaly: Revolutionary Knowledge in Everyday Life

The American political pro-wrestling metaphor becomes all the more interesting when the words of F.A. Hayek, a brilliant polymath and Nobel laureate in economics, sink in. Hayek, with a powerful grasp of analogy and rhetoric, understood and expressed the truth of the choices offered by a 'two-party' system like that of the USA:

> Let me now state what seems to me the decisive objection to any conservatism which deserves to be called such. It is that by its very nature it cannot offer an alternative to the direction in which we are moving. It may succeed by its resistance to current tendencies in slowing down undesirable developments, but, since it does not indicate another direction, it cannot prevent their continuance. It has, for this reason, invariably been the fate of conservatism to be dragged along a path not of its own choosing. The tug of war between conservatives and progressives can only affect the speed, not the direction, of contemporary developments. But, though there is a need for a 'brake on the vehicle of progress,' I personally cannot be content with simply helping to apply the brake. What the liberal must ask, first of all, is not how fast or how far we should move, but where we should move. In fact, he differs much more from the collectivist radical of today than does the conservative. While the last generally holds merely a mild and moderate version of the prejudices of his time, the liberal today must more positively oppose some of the basic conceptions which most conservatives share with the socialists. (Hayek 1960)

When one of the parties in a two party political system is 'conservative,' there is actually only one party, with two different speeds. Conservatism isn't the opposite of socialism; it's just a slowing down of the march toward socialism and toward the centralization of political and financial influence. It is one party with one basic set of policies split into subsets of those who either want change fast (in America, the Democrats) or want it slower (in America, the Republicans). The nation will still arrive at centralized political and financial power; when you look at our dominant paradigm, maybe it already has–the central government, central banks, and centralized media, the McMahons of American politics. Like make-believe pro wrestling matches, the apparent divergent paths that the American votes for is largely make-believe.

Until it was revealed and accepted as theatre and not a *competitive endeavor*, fixed professional wrestling was a fraud. Likewise, manipulated elections presented as fair are fraud. Fraud's rampant throughout the voting processes in America, but it is brushed under the rug. Though taboo, the problems with the fraudulent tallying of votes are well known. A few years ago, HBO aired an excellent documentary called *Hacking Democracy* about the alarming shortcomings and conflicts of interests of the Diebold corporation that manufactured the electronic voting machines used to tally votes in elections nationwide. Most people have at least heard of the terms 'Gerrymandering' and 'ballot stuffing.' The 2004 Bush/Cheney re-election had proved to be rigged; however, it was brushed under the rug as 'incompetence' (M. C. Miller 2006).

This book is not an attempt at scandalum magnatum. It is an attempt to humanize those who are financially, politically, cognitively, and influentially 'distant' from us, to show that the same foibles of egotism, deceit (of self and others), etc. plague all people of all stations in life–prince, pauper, politician, and scientist. Anomaly is the nudity at the parade where the Emperor wears no clothes.

The word conspiracy is derived from Latin *conspirare* (to breathe together); it's simply the idea that two or more people agree to perform an illegal act and keep that act secret. By definition, the Constitutional Convention was a conspiracy. And so was 9/11. But you have a choice in which conspiracy theory you'd like to believe. For example, you can believe the conspiracy that 19 Muslims with box-cutters, led by a super-villain in a cave from the other side of the planet, somehow circumvented the most sophisticated intelligence, defense, and law enforcement system in the history of human kind, or you have the option to believe a conspiracy that has been noted to exist throughout all of recorded history–the conspiracy of the powerful to remain in power and usurp more power surreptitiously (Eggen 2006). Parsimony (that is, Occam's razor) suggests the latter. Again, a Type I error is preferable in the face of imperfect, asymmetrical information; its cost is negligible, while the cost of a Type II error could be catastrophic.

Currently, we are facing a credit-crisis, as the economists say. In financial terms, when you repay your debts you prove that your word is good and are awarded a high credit score. It's your credibility–reputation, trust, and ultimately your penchant for honesty. It then follows that we are in the midst of a 'credibility' crisis or 'crisis of confidence.' For peaceful co-existence, we must be able to trust each other and live up to our contracts (verbal or written) at every class of society, or our interactions will necessitate violence.

The history of human kind is largely the story of slaves and masters. In our earliest records, since at least the time of the Pharaohs, tyranny and cruelty have been the norm and not the exception. The rights offered by the Constitution and the Bill of Rights are anomalies in the paradigm of tyranny. Given the phenomenon of mean reversion, it's exceedingly important to defend the words of the Constitution and Bill of Rights and their application if the norm for the masses is slavery. Our word is our bond, whether it is a vow to your spouse or the vow to "defend the Constitution of the United States against all enemies, foreign and domestic" (Oath of Office).

This crisis of confidence is a problem with the fidelity of second-hand (explicit) knowledge. Whether it is black chambers like the NSA, CIA, MI6, ISI, or powerful special interest groups or wealthy individuals, they have an interest in discrediting those who accuse them of crimes, whether real or imagined. We all have secrets; the powerful are no different, except that their secrets, at times, have a greater impact. In totalitarian regimes, such as the Soviet Union and Nazi Germany, punishment for political dissent went beyond the ridicule of calling those with alternative paradigms 'kooks' or 'crazies.' Frequently, dissent is squelched with more drastic measures, such as a diagnosis of schizophrenia, followed by involuntary commitment and forced sedation.

> *The 'scientific control' of behavior based on the 'science of psychiatry' is a gigantic confidence game which consists of transforming, by means of psychiatric jargon, what is perfectly obvious into what is impenetrably mysterious. Its inevitable result is a series of crimes against humanity, usually perpetuated by the créme de la créme of the profession. In the 1940s and 1950s, some of the most prominent American medical institutions, psychiatrists, and psychologists were engaged in medical 'experiments' that differed only in degree and scope from those engaged in by Nazi physicians experimenting on the inmates of concentration camps. Working in secret for the CIA, psychiatrists systematically poisoned people and used electric shock treatments to destroy their memories, ostensibly in an effort to discover methods of 'mind control'. Like all psychiatric 'abuses' this massive criminal conspiracy against the public was quickly forgotten.* (Szasz 2001, 103)

How do we reconcile anomalies in political economy and the tropes of the mass media? Most that argue against 'conspiracy theory' rightly demand:

1. Predictability: Does the claim accurately and reliably forecast events?

2. Control: Does the claim offer some level of reliable manipulation of things?

3. Falsifiability: Is there enough intension in the claim that it may be falsified?

4. Explanatory power: Does it offer understanding of the way the world works?

Those that argue for a 'conspiracy theory' tend to defend their claims with ad ignorantiam (by definition, those involved in conspiracy are actively hiding or destroying evidence), while those that argue against a 'conspiracy theory' use variants of the fallacy of negative proof (because a premise cannot be proven true, it must be false). Both sides also use ad hominem attacks (calling each other 'close-minded' or 'kooks').

In stark contrast, the anomalist pursues 'conspiracy evidence' eschewing name calling and logical fallacies. It becomes easier to sympathize with the conspiracy theorist mindset by simply extrapolating that our leaders are really no different than we are, that they are not any more or less 'morally superior' than the rest of the population. These men and women have secrets, ethical lapses, and unspoken alliances just as all of us do, the only difference being that their secrets and agendas can adversely affect the health and well being of millions of innocent individuals. As mentioned earlier, 'unintended consequences' seems to be a more popular and accepted explanation to many of our problems, with the conspiratorial approach being looked down upon. Often times, it is impossible to know whether consequences are intentional or unintentional, except for those responsible for the consequences. Conspiracies happen. Our own government has been caught many times, like in the case of Operation Northwoods (a plan by U.S. black chambers to commit apparent acts of terrorism in American cities in order to rally public support for a war against Castro's Cuba) or the inhuman Tuskegee syphilis experiment conducted between 1932 – 1972 (it wasn't until 1997 that President Bill Clinton issued a formal apology to the surviving victims). However, it begs the question; how many times have they not been caught?

Assassinations–Abraham Lincoln, John F. Kennedy, Malcolm X, and Martin Luther King–offer perhaps the best form of conspiracy evidence since the 'unintended consequences' argument doesn't work in these cases. Apart from JFK, Presidents Abraham Lincoln, James A. Garfield, and William McKinley were also assassinated. There are many theories, each with varying amounts of supporting evidence, about what motivated the assassination of JFK. Some say it was because he planned to end the CIA. Others say it was because he planned to end the Federal Reserve. Each case of assassination can be treated as individual instances of conspiracy, or maybe, we can see a pattern. Let's examine the Federal Reserve conspiracy theory further. Abraham Lincoln made the U.S. Treasury issue Greenbacks, making Central Bankers superfluous. Here are Lincoln's words:

> *The money powers prey upon the nation in times of peace and conspire against it in times of adversity. The banking powers are more despotic than a monarchy, more insolent than autocracy, more selfish than bureaucracy. They denounce as public enemies all who question their methods or throw light upon their crimes. I have two great enemies, the Southern Army in front of me and the bankers in the rear. Of the two, the one at my rear is my greatest foe. Corporations have been enthroned, and an era of corruption in high places will follow. The money power of the country will endeavor to prolong its reign by working upon the prejudices of the people until the wealth is aggregated in the hands of a few, and the Republic is destroyed.*

President James A. Garfield was also a known enemy of bankers at the time. He wrote:

> *Whosoever controls the volume of money in any country is absolute master of all industry and commerce...And*

> *when you realise that the entire system is very easily controlled, one way or another, by a few powerful men at the top, you will not have to be told how periods of inflation and depression originate.* (Stabilization of Commodity Prices 1932)

President William McKinley was an out-spoken advocate of the Gold Standard. Two other Presidents, who were against central banking, were Ronald Reagan and Andrew Jackson, and you guessed it, both survived assassination attempts. Ronald Reagan was shot on March 30, 1981. President Jackson's would-be-assassin, Richard Lawrence, gave several reasons for the shooting, one of which was that with the President dead, 'money would be more plenty.' This was no doubt a reference to Jackson's battle with the Bank of the United States.

Are conspiracy theorists merely making Type 1 errors or seeking to integrate anomalies that they notice with regards to governance? Due to the very nature of Black Chambers as secretive agencies, we'll likely never know what is real and unreal with regards to news and history. Sometimes, however, interesting tidbits come out. For example, there are strong reasons to believe that the attack on Pearl Harbor wasn't just a Japanese conspiracy, but perhaps an American one:

> *Republicans were preparing to run Thomas E. Dewey for President. High among their issues was the charge that inexcusable administration laxity had permitted the Japanese attack at Pearl Harbor to succeed so cruelly; there were even hints that President Roosevelt had deliberately invited the attack to get the country into 'his' war over strong isolationist sentiment. Buttressing the charge was the knowledge, circulating secretly among many high officials, that the United States had cracked Japanese codes before Pearl Harbor. From this, many Republicans concluded that the decrypted messages had warned*

Roosevelt of Pearl Harbor and that he, with criminal negligence, had done nothing about it. (Kahn 1967, 604)

Kahn goes even further stating that the "N.S.A. probably owes its existence, like the Central Intelligence Agency and the Department of Defense itself, to Pearl Harbor" (674). One particular anomaly that gets most conspiracy theorists fired-up is MKULTRA, a top-secret mind control and chemical interrogation program run by the Office of Scientific Intelligence, a branch of the CIA.[1]

MKULTRA wasn't brought to the attention of the general public until 1975 when the U.S. Senate committee 'United States Senate Select Committee to Study Governmental Operations with Respect to Intelligence Activities' (aka The Church Commission) chaired by Senator Frank Church was formed due to the crisis of confidence created by the Vietnam Conflict and the ensuing Watergate scandal (due in large part to the aforementioned work of Daniel Ellsberg, as well as that of journalist Seymour Hersch). President Gerald Ford[2] created his own commission, 'U.S. President's

---

[1] You can read about it (under the Freedom of Information Act) by writing CIA. You need to provide your name, address, phone number and date, along with a money order for $30 payable to the US Treasurer, for the request to be processed. The address and format follows:

Central Intelligence Agency
FOIA and Privacy Act Coordinator
Washington, DC 20505

Re: Freedom of Information Act and Privacy Act Request

Pursuant to the provisions of the Freedom of Information and Privacy Acts, 5 USC552 I am formally requesting access to the three-volume CD set containing the entirety the of the Artichoke/Bluebird and MKULTRA records (approx. 20,000 pages).

Enclosed please find my money order for $30 made payable to the United States Treasurer.

Thank you,

[2] Gerald Ford was the only President and the only Vice President that was never actually elected to office. He was appointed by Nixon to the Vice Presi-

ANOMALY: REVOLUTIONARY KNOWLEDGE IN EVERYDAY LIFE

Commission on CIA activities within the United States,' headed by the then Vice President Nelson Rockefeller (whom Ford had nominated to serve as VP upon Nixon's resignation), brother of David Rockefeller, chair of Council of Foreign Relations at the time.[3] David Rockefeller writes:

> *For more than a century, ideological extremists at either end of the political spectrum have seized upon well-publicized incidents such as my encounter with Castro to attack the Rockefeller family for the inordinate influence they claim we wield over American political and economic institutions. Some even believe we are part of a secret cabal working against the best interests of the United States, characterizing my family and me as 'internationalists' and of conspiring with others around the world to build a more integrated global political and economic structure – one world, if you will. If that is the charge, I stand guilty, and I am proud of it.* (Rockefeller 2002, 405)

---

dency, under the terms of the 25th amendment when Spiro Agnew resigned and became President upon Nixon's resignation. Ford then nominated Nelson Rockefeller to serve as his Vice President and to head his investigation of the CIA. Ford was no stranger to Presidential commissions; years earlier, President Lyndon B. Johnson had assigned him to the Warren Commission (which investigated the Kennedy assassination) where he was to prepare a biography of Lee Harvey Oswald. The Assassination Records Review Board was created in 1992 by the U.S. Congress and in 1997 released a document revealing that Ford altered the documents to read, a "bullet had entered the base of the back of [Kennedy's] neck slightly to the right of the spine." According to records found in FBI documents released in 2008, Ford had strong ties to J. Edgar Hoover. This portrait of Gerald Ford is a far cry from the caricature of him performed by Chevy Chase on Saturday Night Live. Ford could very well control what information was made public about the Black Chambers in this way.

[3] It is interesting to note in this context that Nelson and David Rockefeller are the sons and grandsons of the same John D. Rockefeller, Jr. and Sr. eugenicists that created the Rockefeller Foundation mentioned earlier.

The CFR has been the source of many conspiracy theories in its own right:

> *Does it not seem strange to you that these men just happened to be CFR and just happened to be on the Board of Governors of the Federal Reserve, that absolutely controls the money and interest rates of this great country without benefit of Congress? A privately owned organization, the Federal Reserve, which has absolutely nothing to do with the United States of America!* (Goldwater 1979)

> *The CFR is the American Branch of a society which originated in England, and which believes that national boundaries should be obliterated, and a one-world rule established.* (Quigley 1975)

> *The CFR is the establishment. Not only does it have influence and power in key decision-making positions at the highest levels of government to apply pressure from above, but it also finances and uses individuals and groups to bring pressure from below, to justify the high level decisions for converting the U.S. from a sovereign Constitution Republic into a servile member of a one-world dictatorship.* (Democratic Congressmen John R. Rarick)

> *The Council on Foreign Relations, another member of the international complex, financed both by the Rockefeller and Carnegie foundations, overwhelmingly propagandizes the globalist concept.* (Congressman B. Carroll Reece)

Within the Black Chambers and on the periphery of the intelligence community, there are heroes, like Daniel Ellsberg, Ray McGovern, Roger Morris and many others that are willing to challenge the dominant paradigm. Conspiracy theorists make much of the interconnections among the people in power, among their families, and conspiracy-suggesting incidents in their or their family members' lives.[4] The only way we can know whether these connections

---

[4] Here are some interesting linkages: For the Vice Presidency appointment, Gerard Ford's choice came down to Nelson Rockefeller or George Herbert Walker Bush. In 1975, Ford appointed Bush to become Director of the Central Intelligence Agency, a position he held until 1977 when he left to become Director of the Council on Foreign Relations (CFR), which, at the time, was chaired by David Rockefeller.

George H.W. Bush's great-grandfather in law was William Averell Harriman (yes, the son of wealthy eugenicist Mrs. E.H. Harriman). Averell Harriman served as Senior Partner of Brown Brothers Harriman & Co. when the Harriman Bank was the primary Wall Street home for an early financial backer of the Nazi party Fritz Thyssen. In 1942, the Harriman business interests were seized under the Trading with the Enemies Act. This set back, however, didn't stop Harriman from becoming the Secretary of Commerce under Truman and then the Governor of New York, a seat he eventually lost to future VP of the United States, Nelson Rockefeller. As the son-in-law of the wealthy Averell Harriman, Prescott Bush, the grandfather of the 41st and 43rd Presidents of the United States, would soon become a United States Senator. As a senator, Prescott Bush advised both Presidents Eisenhower (the same "fella" that Truman felt "never paid any attention to it [the CIA], and it got out of hand") and Richard Nixon. In fact, Prescott Bush was a major financier of Nixon's presidential campaign against Kennedy.

The Rockefeller/Black Chamber influence within Arkansas politics is also worth noting. Another Rockefeller Oil heir, Winthrop Rockefeller was the 37th Governor of Arkansas, the brother of Vice President Nelson Rockefeller, the father of Winthrop Paul Rockefeller (who, under Governor Mike Huckabee, would become Lieutenant Governor of Arkansas and later make a run for the Presidency of the United States).

Secretary of State Hillary Clinton, a former First Lady of both Arkansas and the United States and a Democratic Presidential candidate in 2008 has also been found to be connected to the affair that put Nelson Rockefeller in the Vice Presidency. In 1973, Jerry Zeifman, the general counsel and chief of staff of the House Judiciary Committee wrote a letter to the New York Post, which was published on August 16th, 1999. In the letter, he claims that Hillary Clinton responsible for 'establishing the legal procedures to be followed in the course of the inquiry and impeachment [of President Nixon]' recommended ethically flawed procedures and violated rules by disclosing confidential information to unauthorized persons.

Hillary Clinton's 'partner in power' Bill Clinton, was a CIA recruit. Roger Morris, a senior staffer on the National Security Council, says in his book *Partners in Power*: "One former agency official would claim that the future president was a are sinister is if a hero, usually an insider, comes forward, filled with sympathy for his fellow human beings, with proof. Until then, we can only guess and speculate about motives. As Marcello Truzzi says "extraordinary claims require extraordinary proof."

For now, the Freedom of Information Act is an extraordinary means provided to ordinary citizens to gather such proof. Every American should contact their Congressperson, their Senators, and dedicated military personnel and elected officials and remind them of their pledge to support and defend the Constitution of the United States against all enemies, foreign and domestic.

In light of the crisis of confidence, people can hardly be blamed for believing that the 'rustle in the grass' might cost them their life. Today, Senator Jay Rockefeller is the chairman of the Senate Intelligence Committee. Yesterday, unelected Vice President Nelson Rockefeller was the chairman of the committee that investigated the assassination of President John F. Kennedy. He was recommended to this position by Henry Kissinger, a man who, as Christopher Hitchens has argued persuasively, is a war-criminal and later led the September 11th Commission at the request of President George W. Bush. It is of the gravest importance that citizens make their representatives keep their pledge and allow anomalies (whether scientific or political) to be discussed and investigated to protect the free flow of ideas and individual's rights, no matter how anomalous or strange the person or idea may be.

> *There is nothing to check the inducements to sacrifice the weaker party or an obnoxious individual. Hence it is that such democracies have ever been spectacles of turbulence and contention; have ever been found incompatible with personal security or the rights of private property; and have in general been as short in their lives as they have been violent in their deaths* (Madison 1961, 81)

---

full-fledged 'asset,' that he was regularly 'debriefed,' and thus that he informed on his American friends in the peace movement in Britain..." (Morris 1996, 103).

CHAPTER 12

THE CENTRALIZATION OF SYMBOLOGY AND THE PARADIGM OF TYRANNY

I spent 33 years and four months in active military service and during that period I spent most of my time as a high class muscle man for Big Business, for Wall Street and the bankers. In short, I was a racketeer, a gangster for capitalism. I helped make Mexico and especially Tampico safe for American oil interests in 1914. I helped make Haiti and Cuba a decent place for the National City Bank boys to collect revenues in. I helped in the raping of half a dozen Central American republics for the benefit of Wall Street. I helped purify Nicaragua for the International Banking House of Brown Brothers in 1902-1912. I brought light to the Dominican Republic for the American sugar interests in 1916. I helped make Honduras right for the American fruit companies in 1903. In China in 1927 I helped see to it that Standard Oil went on its way unmolested. Looking back on it, I might have given Al Capone a few hints. The best he could do was to operate his racket in three districts. I operated on three continents. (Butler 1935)

Major General Smedley Darlington Butler (aka 'The Fighting Quaker' and 'Old Gimlet Eye') was, at the time of his death in 1940, the most decorated Marine in U.S. history. In 1934, he told the McCormack-Dickstein Congressional Commission that he had been approached by a group of wealthy industrialists to lead a military coup to overthrow the government of President Franklin D. Roosevelt. He was offered $3 million to lead half a million angry World War I veterans to physically remove FDR from the White

House to save the American dollar from Roosevelt's New Deal policies. Allegedly, Robert Sterling Singer Clark and J.P. Morgan offered funding, the former providing a $50 million war chest. Remington would supply weapons and DuPont ammunition. Shockingly, many of Butler's claims were confirmed by the McCormack-Dickstein Committee's final report:

> *In the last few weeks of the committee's official life it received evidence showing that certain persons had made an attempt to establish a fascist organization in this country… There is no question that these attempts were discussed, were planned, and might have been placed in execution when and if the financial backers deemed it expedient.* (United States Congress, House of Representatives, House Committee on Un-American Activities 1935)

This unusual episode is unknown to most Americans making me wonder how many others like this have happened that even fewer know about. Some even persuasively claim that the Fascists have indeed succeeded in their paradigmatic coup. Way back in 1975, Charlotte Twight stated in her book *America's Emerging Fascist Economy* :

> *The fascist state initially assumes increased control over a few industries deemed 'vital' to the national interest. The list may vary, but it usually embraces agriculture, armaments (national defense), energy, and finance (as a means of engineering the full spectrum of its economic, social, and political policies). Both Germany and Italy acted early to centralize control over the key areas of agriculture and finance, abrogating capitalism to a degree at that time unusual for the budding fascist economies.* (Twight 1975)

The Centralization Of Symbology And The Paradigm Of Tyranny

Democracy is by its very nature collectivist, placing the welfare of the 'whole' above that of the individual. Popular opinion holds representative democracy to be government that "tends to provide for the interests of the governed and protect them against the abuse of power," and one "in which satisfaction is maximized and conflicts reconciled by pressures bringing countervailing pressures into operation" (The Encyclopedia of Philosophy 1967). Despite its good intentions, representative democracy tends to achieve results antithetical to these intentions. In this chapter, my purpose is to explain how the paradigm of pure democracy tends to actually work against anomalies that may subvert the dominant paradigm and the interests of the individual.

> *Democracy means simply the bludgeoning of the people, by the people, for the people.* (Wilde 2007 (reprint edition), 244)

The United States is a federal constitutional republic; however, our representatives are elected democratically, and politicians are known to break their campaign promises to pursue their own agendas. Individual interests are truly protected only by the constitutional law, all people being equal before the law, no matter how anomalous they are or what they have to say may be. Further, forces that lead to the centralization of symbols, manipulate democracies through the manipulation of money and language:

> *A democracy cannot exist as a permanent form of government. It can only exist until the voters discover that they can vote themselves largesse from the public treasury. From that moment on, the majority always vote for the candidate promising the most benefits from the public treasury, with the result that a democracy always collapses over loose fiscal policy, always followed by a dictatorship.* (Attributed to Sir. Alexander Fraser Tytler)

Constitutional republics usually have mechanisms in place to prevent the majority tyrannizing over dissenting individuals and minority groups. However, the last few elections reveal that, in representative democracies, it is not always the majority that threatens to oppresses the minority (as DeToqueville warned); instead, representative democracy actually allows the tyranny of a minority. In the last presidential election, the majority of Americans did not vote. However, America got its representatives and the President. A narrow and well-organized minority of voters (even discounting the possibility that an influential number of those votes may be fraudulent) determined the outcomes, which the majority that didn't vote had to accept. Either way, if not extremely alert, representative democracy can end up as either De Toqueville's tyranny of the majority or as the Political Action Committee and special interests' tyranny of the minority. Either way, without strong laws protecting the individual, such as the Bill of Rights, you end up with tyranny.

> *They were not Slaves to Tyranny,*
> *Nor ruled by wild Democracy;*
> *But Kings, that could not wrong, because*
> *Their power were circumscrib'd by Laws.*
> (Mandeville 1989, 63)

Many, including Condorcet and Noble laureate Kenneth Arrow, have grappled with fundamental problems with democratic decision-making. These issues still remain unresolved. Condorcet, a brilliant eighteenth century French philosopher and mathematician, explained how the collective decision of a group of rational individuals could produce irrational results. In a small group of individuals where alternative choices are few, the majority rule works well. However, as the number of individuals voting and the number of choices voted upon increase, the performance of 'majority rule' as a barometer of what 'the group' wants decreases (Condorcet 1785). Kenneth Arrow built upon the work of Condorcet to formulate his *Impossibility Theorem* that states that in life, there is an opportunity cost that consists of foregoing a concentration of political power at the expense of social rationality (Arrow 1950).

> *Democracy, at present, defeats its objects by the vastness of the constituencies involved. Suppose you are an American, interested in a Presidential election. If you are a Senator or a Congressman, you can have a considerable influence, but the odds are about 100,000 to 1 that you are neither. If you are a ward politician you can do something. But if you are an ordinary citizen you can only vote. And I do not think there has ever been a Presidential election where one man's abstention would have altered the result. And so you feel as powerless as if you lived under a dictatorship. You are, of course, committing the classical fallacy of the heap, but most people's minds work that way.* (Russell 1952, 72-73)

For centuries, until the recent proliferation of the Internet, the economy for media has favored an information distribution system where a few information engineers (academics, writers, journalists, etc.) feed processed information to a larger number of voting, 'non-information' laborers in other divisions of the economy (manufacturing, retail, etc.). Through licensing and the threat of censorship, the state has bullied information laborers into servitude. Consequently, information is manipulated and filtered causing the voting decisions of the uncritical electorate (the non-information labor numerator in a division of labor society) to advocate whichever pro-statist policy is the pet of the politicians currently in power. Another word for such doctored information is propaganda. The highly influential syndicated columnist Walter Lippman correctly identified the method by which democratic governments retain power–the *manufacture of consent* (Lippman 2003). (For an interesting propaganda model see *Manufacturing Consent: The Political Economy of the Mass Media* by Noam Chomsky and Edward S. Herman.) It requires collusion between the government and the media used to inform the electorate.

> *One of the tendencies of democracy, which Plato and other antidemocrats warned against a long time ago, was the danger that rhetoric would displace or at least overshadow epistemology; that is, the temptation to allow the problem of persuasion to overshadow the problem of knowledge. Democratic societies tend to become more concerned with what people believe than with what is true, to become more concerned with credibility than with truth. All these problems become accentuated in a large-scale democracy like ours, which possesses the apparatus of modern industry.* (Boorstin 1961, 129)

Sadly, innocent students in public schools are the greatest consumers of this engineered information (Richman 1995, 39). Taught by teachers who are extraordinarily reluctant to bite the state hand that feeds them, students are relentlessly indoctrinated with the pro-state ideology. Of all the pro-state propaganda received by these students, the most deceitful is that which tells them that the American economy is a free market one, when in fact it is a fascist one. Fascism, as manifested by Benito Mussolini, was essentially interventionist and advocated policies of a strong nationalistic collectivism.

> *The Culmination of Fascism's Economic Policies: Economic and Psychological Dependence...Economic dependence is fostered on many levels. Businessmen are forced to depend on government licenses as a prerequisite of pursuing their trade, making entrepreneurs subject to bureaucratic whim and governmental fiat for their economic livelihood...As a fascist government increasingly usurps the functions of private enterprise in providing the daily necessities of its citizens such as health care, food,*

> *housing, energy, and insurance, the individual becomes acutely aware that his survival is dependent upon governmental decisions that he as an individual cannot significantly influence...* (Twight 1975)

Let me be clear here that I am not using the term fascism loosely but as an analytical term differentiating one form of political economy from the rest. Alfredo Rocco described Mussolini's *stato corporativo* aptly when he said, *"For Fascism, society is the end, individuals the means, and its whole life consists of using individuals for its social ends"* (Rocco 1926). Democratic, majority rule, fosters this sort of mentality that puts group interests ahead of the individual.

> *Under democratic ideology runs the current of fascism which overflows at the surface.* (Pareto 1964)

Common usage (from my Random House College Dictionary) describes fascism as a form of government that:

1. Exerts control through a centralized state (In the US, federalism with centralized control housed in Washington, D.C.)

2. Customarily fosters racism (read slavery and eugenics movement in the US, apart from less extreme forms of racial bias)

3. Stresses aggressive nationalism (demonstrated amply by events in the US after 9/11, or go to a NASCAR rally to witness it yourself)

4. Is led by a dictator. (Do you need to look any further than Franklin Delano Roosevelt?)

America clearly has become a Fascist state. The only point of contention could be the last criterion. (Note that the Weimar Republic's democratic government was merely a means to Hitler's Nazi ends).

ANOMALY: REVOLUTIONARY KNOWLEDGE IN EVERYDAY LIFE

Roosevelt with Josef Stalin. NOTE: As a dictator, Stalin murdered even more innocents than Hitler...

Despite Roosevelt's oath under Article II, Section I, subsection 8 of the Constitution, to "preserve, protect and defend the Constitution of the United States," he deliberately tried to topple the Madisonian balance of powers outlined in the Constitution by using the executive branch to control the judicial branch. Within a week of taking power, he obstructed Americans from gaining access to their own property by closing banks nationwide (during a bank run caused by the immense credit expansion allowed by fractional reserve banking endorsed by the Federal Reserve) and called a special session of Congress known as 'The Hundred Days' which granted him immense control over the economy through such legislation as the *Tennessee Valley Authority Act,* the *Agricultural Adjustment Act,* the *National Industrial Recovery Act,* and the *Emergency Banking Relief Act of 1933,* which completely by itself suspended the right of American citizens to own property, allowing government to seize any gold held privately. According to the rhetoric of the day, you weren't a good American if you selfishly 'hoarded' what rightly belonged to you.

> *Fascism's Philosophic Core: Capitalistic Collectivism...Since private property and the profit motive work effectively as incentives for high production, fascism uses these features of capitalism insofar as they do not conflict with the national interest as formulated by fascism's political authorities.* (Twight 1975)

When the Supreme Court deemed the National Recovery Administration, the Agricultural Adjustment Act, and a host of Roosevelt's other New Deal policies blatantly unconstitutional, F.D.R responded by packing the Supreme Court with justices who would not oppose his ascent to dictatorship. By 1944, all but two were justices appointed by Roosevelt himself. The mergers and acquisitions in the media that serve the disinformation campaigns produced by Black Chambers, the dominance of the Executive Branch (or the 'Cult of the Presidency' as one author calls it), and the corruption and collusion between corporations and governments, all grant incredible power over our lives to just a handful of people.

Roosevelt's New Deal policies were followed by the Japanese-American evacuation and relocation in World War II, which began as a 'voluntary program.' When it became clear that Japanese-American citizens did not really want to relocate to internment camps, President Roosevelt issued *Executive Order 9102* and the War Relocation Authority was created. More than 110,000 Japanese-Americans were made to move from their homes **under threat of physical violence from the state** and to surrender their property and jobs, and some were imprisoned in the internment camps for up to three years! (Smith 1995)

> *A further, more specific reason why, in 1933, the New Deal was often compared with Fascism was that with*

> *the help of a massive propaganda campaign, Italy had several years earlier begun the transition from a liberal free-market system to a state-run or corporatist one. In the 1930s, corporatism was increasingly regarded internationally as a perfectly comprehensive response to the collapse of the liberal, free market economy–as was the policy of national self-sufficiency practiced by the Stalinist Soviet Union in withdrawing from the world economy. Of course, the Italian corporatist program–which historian Maurizio Vaudagna calls 'Fascism's most original innovation up to that point'–seemed infinitely preferable to the Communist 'great leap forward' because it didn't involve the expropriation of private property. There was hardly a commentator who failed to see elements of Italian corporatism in Roosevelt's managed economy under the National Recovery Administration, the institution formed in 1933 to maintain mandatory production and price 'codes' for American industry. The Italian press was quite taken with these similarities, and Mussolini laid the groundwork for such comparisons in a book review he wrote of Roosevelt's Looking Forward. ...he identified a spiritual kinship:*
>
> *The appeal to the decisiveness and masculine sobriety of the nation's youth, with which Roosevelt here calls his readers to battle, is reminiscent of the ways and means by which Fascism awakened the Italian people.* (Schivelbusch 2006, 23)

Before Roosevelt, however, Woodrow Wilson paved the way for fascism in America. Wilson brought a lot of the beginnings of a fascist state to America during his administration with:

1. Entangling America in World War I through propaganda, which directly led to the rise of Hitler, Lenin, Stalin, and the events of World War II.

The Centralization Of Symbology And The Paradigm Of Tyranny

2. Creating the League of Nations (a first attempt at one-world-government and forerunner of the United Nations)

3. Racism. Here is his quote from the racist film *Birth of a Nation*:

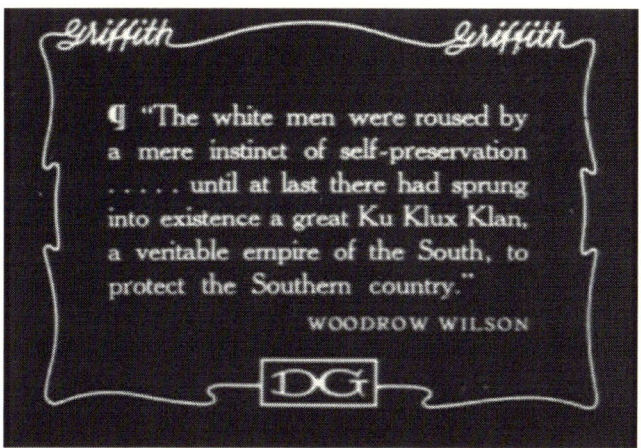

The concentration of state-power in the executive branch and influence (in the form of propaganda) were not the only forms of centralization institutionalized by Woodrow Wilson. He also centralized financial and economic power with the creation of the Internal Revenue Service (a part of his reform package that passed Congress) and the implementation of the quasi-government system of credit Federal Reserve in 1913.

> *If the American people ever allow private banks to control the issue of their currency, first by inflation and then by deflation, the banks and corporations that will grow up around them will deprive the people of all property until their children will wake up homeless on the continent their fathers conquered.* (Jefferson, Letter to Albert Gallatin, Secretary of the Treasury 1802)

The centralization of credit under the auspices of government has had as much of a stifling effect, if not more, as the centralization of influence under the institutionalization of Wilson and Roosevelt's propaganda systems here in the United States. Again, the Bismarckian influence of centralization and *machpolitik* was not just permeating education and medicine, it was infecting finance and economics, as well. As early as 1906, the New York Chamber of Commerce recommended the creation of a centralized banking system patterned after the Reichsbank. Just one year later, the Aldrich Bill was passed which allowed, but did not mandate, central banking activity by the Treasury Department of the United States. In September of 1909, both George Reynolds, president of the American Bankers Association, and William Howard Taft, President of the United States, publically stumped for the establishment of a centralized banking system modeled upon the German Reichsbank.

In 1910, Senator Nelson W. Aldrich (father-in-law to John D. Rockefeller, Jr., and grandfather of Nelson Aldrich Rockefeller, the Vice President of the United States under Gerald Ford) and the Assistant Secretary of the Treasury Department A. Piatt Andrews invited Frank A. Vanderlip of the National City Bank, Paul Warburg of the wealthy German Warburg banking family, and J.P. Morgan partner Henry Davison to a secret meeting (read conspiracy) on Jekyll Island, Georgia, at a private resort owned by John D. Rockefeller and J.P. Morgan to develop a strategy to implement the centralized banking system. "To guard against revelations of their identity and their purpose, they took elaborate precautions, traveling separately to Hoboken, where they boarded a private railroad car for Savannah, using only their first names in front of the train crew" (Federal Reserve Bank of St. Louis). In 1913, Woodrow Wilson signed the Federal Reserve Act into law, effectively centralizing the system of banking and credit in the United States. Warburg would eventually run Kuhn, Loeb, and Co., which was the leading investment house for John D. Rockefeller.

The Centralization Of Symbology And The Paradigm Of Tyranny

> *We have restricted credit, we have restricted opportunity, we have controlled development, and we have come to be one of the worst ruled, one of the most completely controlled and dominated Governments in the civilized world no longer a Government by free opinion, no longer a Government by conviction and the vote of the majority, but a Government by the opinion and duress of a small group of dominant men.* (Wilson 1913)

Just as the Flexner report replicated the German (then Prussian) medical system here with disastrous results, so has the adoption of the Reichsbank centralized bank led us to the current state of our own pervasive financial iatrogenesis. Trope has led to the atrocities of iatrogenesis and the human rights violations of psychiatry, and now, trope is running roughshod over our system of banking and credit causing the largest economic disaster in modern history. Currency inflation, the debasing of money supply, is trope, plain and simple. As Dr. Thomas Szasz writes:

> *Orderly human relations depend on the proper functioning of speech, speakers and listeners attaching the same meaning to words. The languages we speak and how well or poorly we speak them define who we are and largely determine who the persons near and dear to us are. Words have standardized meanings...Safeguarding the fixity of the meaning of words (and other symbols) is essential for the integrity and pursuit of the sciences and is indispensable for law, economics, commerce, and honest dealing among upright persons. Conversely, corrupting the meaning of words undermines their integrity, obstructs cultural and scientific progress, and hinders honest discourse among people.*
>
> *Anchoring money in an objective standard (gold) serves the interests of free trade, the security of property, and personal liberty. Anchoring diagnosis in an objective*

> *standard (somatic pathology) serves the interests of medical science, sound medical practice, and personal liberty. Dislodging the meaning of these symbols from their precisely defined positions–legitimizing fiat money and fiat diagnosis as 'real' money and real diagnosis–serves the interests of the new political class, the pharmacrats.*
>
> *As long as the gold standard was the accepted measure of money, no abstract monetary standard, or 'numeraire' was needed; similarly, as long as the gold standard of disease was the accepted measure of disease, no abstract disease standard, or 'diagnostic numeraire,' was needed.* (Szasz 2001, 53)

To better comprehend the implications, we must understand the relationship between the United States Treasury and the Federal Reserve System. For the United States, the last century has been one of deficits and debt. In simple terms, a deficit occurs whenever you spend more than you have. So every time the federal government spends more than it has, it must issue a debt instrument, or I.O.U., usually a U.S. Treasury bond, to cover the expense. The Federal Reserve banking cartel buys these bonds (with paper currency literally created out of thin-air) on the promise that the government will pay the Federal Reserve back, both the principal and a fixed rate of interest. In exchange for this ongoing cash flow in the form of interest payment, the Federal Reserve literally creates money (mostly electronically and completely out of thin air) through manipulated ledger accounts and gives it to the Treasury. How does the Treasury generate revenue to pay off its debt to the Federal Reserve? The main way is through taxation. Simply put, our income tax goes directly to the central bankers. The 16th Amendment essentially made the American people, via the Internal Revenue Service, surety for the debts of the central bank.

Eventually, this fiat money borrowed from the Federal Reserve trickles into the economy as the government spends it and finds

The Centralization Of Symbology And The Paradigm Of Tyranny

its way back into the private banks. Once there, the real inflation begins through the magic of fractional reserve banking. In a nutshell, since, by law, the banks must only maintain a fraction of the actual reserves on-hand (while their ledgers falsely say they have the whole amount), the currency is inflated and the risk of bank runs remains perennial. This inflationary process is documented in the Federal Reserves' own manual *Modern Money Mechanics*, originally published by the Federal Reserve Bank of Chicago:

> *The total amount of expansion that can take place is illustrated on page 11. Carried through to theoretical limits, the initial $10,000 of reserves distributed within the banking system gives rise to an expansion of $90,000 in bank credit (loans and investments) and supports a total of $100,000 in new deposits under a 10 percent reserve requirement.* (Federal Reserve Bank of Chicago 1971, 8)

Even more sobering is the extent of revenue the IRS must collect to pay the Treasury's bondholders (that is, the Federal Reserve banking cartel and foreign governments). The National Debt towers at nearly $12 trillion (remember a trillion is a thousand billion, and a billion is a thousand million, and million is one-thousand-thousand). At an estimated population of 306,902,359, each US citizen's share of the outstanding public debt is nearly $40K.[1] This is the money that is owed to the Federal Reserve (which just bought, in early 2009, another $300 billion by creating money out of thin air) and foreign governments (the top five being China, Japan, United Kingdom, Oil Exporters, and Caribbean Banking Centers according to http://www.treas.gov/tic/mfh.txt).

> *For this intellectual shell game, the cartelists needed the support of the nation's intellectuals, the class of professional opinion-molders in society. The Morgans needed*

---

[1] http://www.brillig.com/debt_clock/

> *a smokescreen of ideology, setting forth the rationale and the apologetics for the New Order. Again, fortunately for them, the intellectuals were ready and eager for the new alliance. The enormous growth of intellectuals, academics, social scientists, technocrats, engineers, social workers, physicians, and occupational 'guilds' of all types in the late nineteenth century led most of these groups to organize for a far greater share of the pie than they could possibly achieve on the free market. These intellectuals needed the State to license, restrict, and cartelize their occupations, so as to raise the incomes for the fortunate people already in these fields. In return for their serving as apologists for the new statism, the State was prepared to offer not only cartelized occupations, but also ever-increasing and cushier jobs in the bureaucracy to plan and propagandize for the newly statized society. And the intellectuals were ready for it, having learned in graduate schools in Germany the glories of statism and organicist socialism, of a harmonious 'middle way' between dog-eat-dog laissez-faire on the one hand and proletarian Marxism on the other. Instead, big government, staffed by intellectuals and technocrats, steered by big business and aided by unions organizing a subservient labor force, would impose a cooperative commonwealth for the alleged benefit of all.* (Rothbard 1999, 4)

The tricky question is what if the growth of the debt remains constant and greater than the rate of growth of average real income (which is what is taxed)? What then should we expect the government to do when tax revenues are no longer sufficient to pay the interest on the debt? Repudiation is not a viable option given the interest and influence of the politically and financially powerful. Nor could it result in hyperinflation, by the same logic. If you think with your *cui bono* hat on, the mostly likely scenario is one of massive deflation, as that would benefit the bankers most by increasing the value of their assets (including the loans, with interest due, that they've made). While all this is speculation, I'd lay the odds, at this writing, of repudiation at 5%, hyperinflation at 30%, and deflation at 65%.

The case for a *cui bono* deflation scenario is compelling. The U.S. government is mostly borrowing its way to solvency, not printing it via the Federal Reserve. The U.S. Treasury Department auctions are setting records for sales of federal bonds and T-bills (Treasury Bond Auctions Show Insatiable Debt Demand 2009). These sales records are being set despite the interest paid on these debt instruments being at record lows. This may sound strange until you realize that the insatiable demand for this U.S. debt is not driven by the interest paid on them.

In a deflationary scenario, all you have to do is hold the debt to make money; interest isn't needed to make money since as prices fall during a deflation, each dollar buys more than it could earlier. Granted, the Federal Reserve is printing money, but it won't 'trickle down' since the banks are using the stimulus money to shore up their capital reserves due to the trillions of dollars of bad loans they made and the disastrous credit derivatives positions they hold. If the banks were to use mark-to-market accounting and actually show their real market value on balance sheets, they will be shown to be insolvent.

As shown earlier, when the fiat money created from the Federal Reserve reaches private banks, inflation comes as loans to businesses and individuals through fractional reserve banking. But when these loans go bad and individuals and businesses default on their obligations to pay back the loan, that money is simply destroyed. Loan defaults are not inflationary, they are deflationary and loan default rates are at an all time high.

The only thing that's near 100% certain is the continued centralization of power, wealth, and influence. Perhaps, this will manifest itself as a single world currency. The United Nations has already suggested this in the controversial report, the *UN Conference on Trade and Development.* In an echo of the events that culminated at Jekyll Island, there seems to be afoot a push for a new 'Bretton Woods-style' system of managed international exchange rates where the central banks would manipulate their systems of credit based upon the behavior of the rest of the world economy. Reuters in September, 2009, reported in Pittsburgh, Pennsylvania, the march towards centralization as "New world economic order takes shape at G20" (Wroughton 2009).

The centralization of wealth has been illustrated recently by physicists James Glattfelder and Stefano Battiston in "Backbone of complex networks of corporations: The flow of control" (Glattfelder and Battiston 2009). They estimate that world's stocks are actually controlled by a very select few, the most influential of which is the Capital Group Companies started by Jonathan Bell Lovelace back in 1931 (it was called Capital Research and Management Company back then). It is the single most powerful controlling shareholder on the planet with significant control of the capital in 32 of 48 countries considered. This information is not a secret, though little known, and there is nothing nefarious about the Capital Group Companies. It seems like a great corporation. However, the extent of their control of capital is an illustration of this tendency toward centralization.

Many intellectuals, even some of my heroes like Bertrand Russell and Albert Einstein, have promoted the idea that the ultimate centralization, in the form world government, is the only means to peace in the nuclear age and, hence, the only means to human survival. This seems like a false dilemma since federalism is not a guarantee of peace, as the American Civil War can attest. Centralizing all the power, wealth, and influence into even fewer individuals seems like a recipe for further disaster, given the awful record of centralization for human rights. In fact, the decentralization of wealth, the decentralization of power, and the decentralization of influence seem as important as they are unusual. Human history is largely the history of small groups of powerful elites tyrannizing and enslaving large portions of the population for selfish benefit of themselves or their race or 'bloodlines.' As free beings and to continue as free beings, it is of utmost importance to protect the ultimate law of America from the dangers of trope and, thus, prevent what statisticians call a 'mean-reversion' or a trend backwards toward slavery. **Transparency**, not secrecy, and **decentralization**, not centralization, of language, power, and resources alone can prevent tragic history from repeating itself.

> *I wouldn't go to war again as I have done to protect some lousy investment of the bankers. There are only two things we should fight for. One is the defense of our homes and the other is the Bill of Rights. War for any other reason is simply a racket.* (Butler 1935)

CHAPTER 13

CHANCE AND FINANCIAL ANOMALIES

Western philosophy begins with Thales. (Russell 1945)

Thales was a pre-Socratic philosopher from the ancient Greek city of Miletus. Aristotle hailed Thales as the founder of the school of natural philosophy–the first person to investigate the true fundamentals of matter. Not one to be trifled with, when ridiculed by his detractors that his poverty was proof that his philosophy was useless and impractical, Thales set out to prove them wrong.

Being particularly good at weather forecasting, Thales predicted the next fall olive harvest to be particularly exceptional. With this in mind, Thales practically invented, in Aristotle's words, "a financial device, which involves a principle of universal application." Thales invented the first recorded 'option contract.' He negotiated an inexpensive deposit on every olive press in the area to guarantee his exclusive use during the coming harvest season. The press owners were thrilled because they got his deposit money even if the harvest was abysmal and there was no demand for olive presses. As it turned out, his prediction was right and the harvest was bountiful. Thales then substantially raised the price for the use of olive presses since he now effectively held a monopoly on the process of producing olive oil in the region. Had the harvest failed to produce, Thales would have limited his losses to only the original deposit paid. This form of insurance that Thales used is today called an 'option,' a financial instrument that is included in the financial derivative family.

A financial derivative is called 'derivative' because its value is derived from an underlying asset, such as a stock or a commodity. Stock options, oil futures, and interest rate swaps are some of the more commonly known financial derivatives. These financial derivatives are largely the domain of quantitative finance. Practitioners are usually called 'financial engineers' or 'quants' and tend to come from math and physics PhD programs although there has been an explosion of graduate programs in financial engineering since the mid-00s (I received my Master of Science designation in financial engineering in 2002).

The next massively significant figure in quantitative finance after Thales was Louis Bachelier, a French mathematician, at the turn of the twentieth century. His dissertation *The Theory of Speculation*, published in 1900, introduced Brownian motion as a random walk model of stock price changes and demonstrated its application to option pricing (Bachelier 1900). This concept did not prove very contagious until Harry Markowitz began to formulate the theory of optimal portfolio selection in the context of trade-offs between risk and return in 1952. Markowitz's Modern Portfolio Theory (MPT) marked the founding of modern quantitative finance with the idea of portfolio diversification as a risk-reducing strategy (Markowitz 1952). He reformulated, in mathematical terms, the old adage 'don't put all your eggs in one basket.'

> *Thus, Markowitz and others transformed investing from a game of stock tips and hunches to an engineering of means, variances, and 'risk aversion' indices. In fact, the term 'financial engineering' has been popular on Wall Street ever since.* (Mandelbrot 2004, 65-66)

In 1965, just over a decade after Markowitz's MPT was introduced, Eugene Fama published his dissertation 'The Behavior of Stock Market Prices' which updated Bachelier's random walk hypothesis. Thus, was born Fama's Efficient Market Hypothesis (EMH) (E. Fama 1965).

> *It is more than a metaphor to describe the price system as a kind of machinery, or a system of telecommunications which enables individual producers to watch merely the*

> *movement of a few pointers, as an engineer might watch the hands of a few dials, in order to adjust their activities to changes of which they may never know more than is reflected in the price movement.* (Hayek 1948)

Though Hayek's rhetorical flourish that the price system is 'more than metaphor' may be an exaggeration, it's still a very powerful metaphor. According to EMH, in a perfect market, all the relevant information is already priced into traded assets. The efficient market hypothesis is quite appealing conceptually and empirically, which accounts for its enduring popularity. In a nutshell, efficient stock markets are generally thought of as equilibrium markets in which security prices fully reflect all relevant information that is available about the 'fundamental' value of the securities. So persuasive is the rhetoric of the efficient market hypothesis that Benjamin Graham, famous for co-authoring the fundamentalist treatise *Security Analysis* with David L. Dodd, was quoted as saying shortly before his death, "I am no longer an advocate of elaborate techniques of security analysis in order to find superior value opportunities…I doubt whether such extensive efforts will generate sufficiently superior selections to justify their costs…I'm on the side of the 'efficient market' school of thought." (Malkiel 1996, 191)

There is a subtle but very important distinction between what Hayek says and what is claimed by the EMH. EMH says all relevant information, whereas Hayek says "which they may never know more than is reflected in the price movement." Unlike EMH, Hayek takes into account the asymmetrical nature of information and the limits of knowledge–the fact that paradigms and anomalies exist.

Before we continue along our tour of the history of quantitative finance, it is important to know a few anomalies present in the EMH. Despite its popularity, efficient capital markets theory has weathered some very appropriate criticisms. Since a financial theory is still only a model of reality and not reality itself, anomalies arise where theory does not match reality and the theory of efficient capital markets is no exception. Ray Ball's article *The Theory of Stock Market Efficiency: Accomplishments and Limitations* (1994) illuminates some interesting anomalies:

1. Prices overreact to new information, which is then followed by a correction, allowing contrarian investors to take profits (suggested by Kenneth R. French and Richard Roll (1986)).

2. 'Extraordinary delusions and madness of crowds' result in excess volatility of prices.

3. Prices underreact to quarterly earnings reports.

4. There is no relationship between historical betas and historical returns which has led many to believe the equilibrium-based Capital Asset Pricing Model–developed greatly due to the enormous amount of empirical data on efficiency–has failed (Fama and French 1988).

5. Seasonal patterns are found in the data on stock returns of small firms, such as the 'January effect,' where stock prices are unusually higher during the first few days of January, or the 'weekend effect,' where stock returns on Mondays are often quite a bit lower than those of the Friday just before.

There are a few anomalies that Ball's article does not talk about. There's evidence that:

1. Firms with low price-earnings ratios outperform those with higher price-earnings ratios (Dreman and Berry 1992).

2. Stocks that sell with low book-value ratios tend to provide higher returns.

3. Stocks with high initial dividends tend to provide higher returns (Malkiel 1996, 204-207).

Ball also discusses the general neglect of the processing and acquisition costs of information within the theoretical and empirical research on stock market efficiency. This neglect could be the reason for the anomalies, such as the 'small firm effect' and the tendency of small cap stocks to provide higher returns. He also criticizes the assumption in the efficient markets hypothesis of investor 'homogeneity' and suggests the need for a new research program (Ball 1994, 41-46).

Ball ends his piece with the rhetorical question "Is 'behavioral' finance the answer?" followed quickly by, "I don't think so" (Ball 1994, 47). However, 'behavioral' finance could yield useful answers. That investors behave rationally, that is, investors accurately maximize expected utility is an important assumption of the efficient market hypothesis, and if it is not true, it may explain why the anomalies exist. Most models in finance and economics begin with the basic assumption of *Homo Economus*, that man is a rational being. Work in prospect theory by Allias, Kahneman, and Tversky provides important evidence that the standard assumption of expected utility maximization assumed by most financial economists may not furnish accurate representations of human behavior (the prospect theory states that individuals are better represented as maximizing a weighted sum of 'utilities,' determined by a function of true probabilities which gives zero weight to extremely low probabilities and a weight of one to extremely high probabilities). While such evidence is not damning, it is troubling to say the least (Shiller 1997).

Coming back to our tour of the big ideas in quantitative finance, after Fama's EMH, the next big development came in 1973. The Black-Scholes option pricing model ended the long search for a formula by assuming that the option price is independent of both the risk preferences of investors and the expected return of the underlying asset. Five years later, Cox, Ross, and Rubinstein would develop a binary model as a discrete time alternative to pricing options in place of the continuous Black-Scholes model. In the 1990s, quantitative finance would see the development of Value-at-Risk (VaR), a controversial method of assessing risk that aimed to help executives answer the question: What is the probability of not losing more than some certain amount within a certain amount of time in the near future? The common theme across the history of quantitative finance is the recognition of risk with regards to the investment decision-making process.

> *In the end, a theory is accepted not because it is confirmed by conventional empirical tests, but because researchers persuade one another that the theory is correct and relevant.* (Fischer Black; cited in Derman (2003))

In finance, risk is usually modeled in terms of the statistical concept of a standard deviation. The standard deviation is simply the average deviation from our expectation, and it quantifies the amount of unpredictability about a particular outcome.

Most times finance is concerned about the risk of asset prices. These asset prices are assumed to follow a stochastic process called geometric Brownian motion. A stochastic process is a sequence of random variables in time, such as the change in temperature outside right now. When you look at the zig-zag line of an asset's price over time, it appears to be random, or stochastic. Geometric Brownian motion was chosen as the metaphor for the change in an asset's price over time since the price seems to move randomly, unpredictably.

Now, the price of an asset can only be a positive number. This is why when price changes are modeled, which can be negative or positive, a normal distribution is used. However, when the prices themselves are modeled (not the change in price) a lognormal distribution is used because in a lognormal distribution only positive values have a non-zero probability of occurring. In this same way, simple Brownian motion isn't used when modeling an asset price since Brownian motion can be negative. Geometric Brownian motion makes use of the probability distribution of a random variable whose logarithm is normally distributed. Geometric Brownian motion is lognormally distributed, so it seemed to make sense to use the geometric variety to model the random nature of asset prices over time.

However, the devil (as always) is in the details. Let's look at some important trope that occurs when Brownian motion is used as an analogy to describe the randomness of asset prices. Brownian motion assumes that changes are statistically independent. This means that past price changes provide no information about future price changes. If the Brownian motion has been moving down lately, there is absolutely no reason to believe that it will continue to go down (or start going up for that matter). Unfortunately, this is not true. In the same 1988 study by Fama and French, they found a measurable tendency for a stock doing well in one decade to do worse in the next. Their 1986 study also noted price correction, a clear demonstration that prices are not independent.

Brownian motion process is also a martingale, which means that our best guess of a value (that is subject to the process of Brownian motion) is its current value. That is to say that at any one time, the current price fully represents all the information. The martingale assumption doesn't seem to demonstrate any anomalies, so we can let that one slide. The last assumption is extraordinary though. In financial markets, we observe very large movements much more frequently than would be predicted by a normal 'Gaussian' distribution. Unfortunately, the metaphors of finance are not very close to reality.

Brownian motion assumes that asset returns are unpredictable between any two points in time, and thus, they can only be described in terms of a random process that produces a probability of distribution prices. Unfortunately, this distribution is assumed to take the form of a normal, bell-shaped curve. It seems that just as the changes in asset prices are not continuous, but jump about wildly so is it that price changes are not normally distributed. As mathematician and the Father of Fractal Geometry Benoit Mandelbrot notes:

> So on August 4, the Dow Jones Industrial Average fell 3.5 percent. Three weeks later, as news from Moscow worsened, stocks fell again, by 4.4 percent. And then again, on August 31, by 6.8 percent...
>
> The standard theories, as taught in business schools around the world, would estimate the odds of that final, August 31, collapse at one in 20 million–an event that, if you traded daily for nearly 100,000 years, you would not expect to see even once. The odds of getting three such declines in the same month were even more minute: about one in 500 billion. Surely August had been supremely bad luck, a freak accident, an 'act of God' no one could have predicted. In the language of statistics, it was an 'outlier' far, far, far from the normal expectation of stock trading.
>
> Or was it? The seemingly improbable happens all the time in financial markets. A year earlier, the Dow had fallen 7.7 percent in one day. (Probability: one in 50 million) In July 2002, the index recorded three steep falls within seven trading days. (Probability: one in four trillion) And on October 19, 1987, the worst day of trading in at least a century, the index fell 29.2 percent. The probability of that happening, based on the standard reckoning of financial theorists, was less than 10^5–odds so small they have no meaning. It is a number outside the scale of nature. You could span the powers of ten from the smallest subatomic particle to the breadth of the measurable universe–and still never meet such a number.

(Mandelbrot 2004, 3-4)

The anomaly that Mandelbrot makes note of is a huge and glaring one; price changes modeled using a Gaussian 'normal' distribution will routinely overlook 'outliers.' An 'outlier' is an anomaly. It lives in the tails of distributions, tails that in financial markets should be much fatter (or exhibit high *kurtosis* if you are looking to impress your statistician friends).

There are many benefits to financial engineering with regards to insuring losses like Thales did. For example, you can hedge losses in a falling market by shorting index futures, or in a volatile market, you could set up a profit strategy using a 'straddle' (or a similar 'strangle') combination of options contracts or a 'collar' (which allows an investor to limit their gains/losses in volatile times). Unfortunately, there are also great dangers presented by the current paradigms in quantitative finance. The anomalies and paradigms mentioned in this chapter have immediate and profound effects on our everyday lives. Warren Buffett, one of the richest investors in the world has called derivatives "financial weapons of mass destruction," and rightly so. But the danger is not necessarily just the derivatives; it is also the models of quantitative finance and complicated accounting schemes (that claim banks don't have to book current losses from derivatives) that are perhaps even more dangerous.

Quantitative finance has played crucial roles in many of the greatest financial disasters in modern history–the Enron fiasco, the Long Term Capital Management bailout, municipal bankruptcies (like the one in Orange County, California, in 1994), the need to recapitalize re-insurer AIG due to Credit-Default Swaps positions. AIG was nationalized precisely because it couldn't continue in the business of insuring $300 trillion in derivatives. In many ways, our current predicament in capital markets can be attributed to the tropes of financial engineering:

> *Models are a helpful way of looking at the world. If you can get everyone to look at the world your way, then you can sell them things based on your views. This isn't dishonest. It's a reflection of the fact that the locus of financial value is vague and confusing, and any order you*

can plausibly impose on prices is immensely helpful to investors. Unless you can replicate perfectly and hold to expiry, a large part of value is in the mind. (Derman 2003)

CHAPTER 14

THE ANOMALIST

The absence of proof is not proof of absence. –William Cowper (1731-1800)

Knowledge changes over time. Ptolemaic cosmology gave way to the Copernican view of the universe. Lamarck's worldview overtook the Creationist's view of biology, which, in turn, was replaced by Darwin's. Quantum mechanics has superseded Newton's classical mechanics. The cognitive revolution in psychology has left the behaviorists in the dust. Every time, a new model, offering more intension, control, predictability and falsifiability replaces the old, less scientific one. At the heart of these revolutions is anomalistics.

Anomalistics has two central features. First, its concerns are purely scientific. It deals only with empirical claims of the extraordinary and is not concerned with alleged metaphysical, theological or supernatural phenomena. As such, it insists on the testability of claims (including both verifiability and falsifiability), seeks parsimonious explanations, places the burden of proof on the claimant, and expects evidence of a claim to be commensurate with its degree of extraordinariness (anomalousness). Though it recognizes that unexplained phenomena exist, it does not presume these are unexplainable but seeks to discover old or to develop new appropriate scientific explanations.

As a scientific enterprise, anomalistics is normatively skeptical and demands inquiry prior to judgement, but skepticism means doubt rather than denial (which is itself a claim, a negative one, for which science also demands proof). Though claims without adequate evidence are

> *usually unproved, this is not confused with evidence of disproof. As methodologists have noted, an absence of evidence does not constitute evidence of absence. Since science must remain an open system capable of modification with new evidence, anomalistics seeks to keep the door ajar even for the most radical claimants willing to engage in scientific discourse. This approach recognizes the need to avoid both the Type I error - thinking something special is happening when it really is not - and the Type II error - thinking nothing special is happening when something special, perhaps rare, actually occurs (Truzzi, 1979a and 1981). While recognizing that a legitimate anomaly may constitute a crisis for conventional theories in science, anomalistics also sees them as an opportunity for progressive change in science. Thus, anomalies are viewed not as nuisances but as welcome discoveries that may lead to the expansion of our scientific understanding (Truzzi, 1979b).*
>
> *Anomalists search for patterns in the acceptance and rejection of new scientific ideas, and this may involve the history, sociology, and psychology of science as well as the scientific fields themselves.* (Truzzi 1998)

The anomalist tries to avoid both Type I and II errors, errors in deductive reasoning, as well as informal fallacies and cognitive biases. It is this balance between dismissing nonsense while being open-minded that the anomalist constantly strives for. The thin line separating protoscience from scientism (that is, pseudoscience) is where the anomalists find themselves drawn to. Marcello Truzzi, a former professor of sociology at Eastern Michigan University, has written extensively on the anomalistic perspective.

> *In science, the burden of proof falls upon the claimant; and the more extraordinary a claim, the heavier is the burden of proof demanded. The true skeptic takes an*

> *agnostic position, one that says the claim is not proved rather than disproved. He asserts that the claimant has not borne the burden of proof and that science must continue to build its cognitive map of reality without incorporating the extraordinary claim as a new "fact." Since the true skeptic does not assert a claim, he has no burden to prove anything. He just goes on using the established theories of "conventional science" as usual. But if a critic asserts that there is evidence for disproof, that he has a negative hypothesis –saying, for instance, that a seeming psi result was actually due to an artifact–he is making a claim and therefore also has to bear a burden of proof.* (Truzzi 1987, 3-4)

This echoes Wittgenstein's admonishment that "[w]hereof one cannot speak, thereof one must be silent." Sadly, today new ideas (perhaps revolutionary ones) are often subject to ridicule, ad hominem attacks, and scientism, instead of being granted a respectful agnostic silence. Type I errors are commonly debunked (which is great), while Type II errors are mostly ignored or ridiculed, ultimately remaining uninvestigated (which is not so great).

> *The weight of evidence for an extraordinary claim must be proportioned to its strangeness.* (Pierre-Simon, Marquis de Laplace)

This conflict between radical doubters (that is, modern 'skeptics') and open-minded critical thinkers (that is, anomalists) has been documented as far back as Sextus Empiricus (c. 160-210 AD). Sextus Empiricus was an Ancient Greek philosopher and physician whose philosophical works provide the modern reader with the most comprehensive account of ancient Greek and Roman skepticism. These works often had provocative titles, such as *Against the Professors* and *Against the Mathematicians*. Sextus Empiricus questioned the validity of inductive reasoning centuries before the problem of induction was presented by David Hume (Empiricus 1933, 283).

For Sextus, there are three basic types of philosophers–dogmatists, academics, and skeptics. Dogmatists argue that they know the truth, academics argue that such truth cannot be known, and skeptics suspend judgment while continuing to investigate. In this model of skepticism, one is agnostic to the claim to truth. The ideas recorded in Sextus Empiricus's books are ascribed to Aenesidemus and Pyrrho of Elis (c. 360 to c. 270 BCE) and suggest that the skeptic enjoys tranquility as one who by suspending judgment determines nothing. The anomalist is a Pyrrhonian Skeptic and is particularly skeptical of the modern or 'academic' skeptic who may be hastily dismissive of unusual claims.

It is this spirit that fostered the successes of the likes of Robert Ripley and Charles Fort in the early twentieth century. The 'Ripley's Believe It or Not' cartoon strip started it's very own cottage industry of books, television shows, and museums here in America. Charles Fort promoted his magazine *The Fortean Times* upon the strength of the demand for the unusual. In 1919, his book *The Book of the Damned* promoted the idea that social values (what Kuhn would later call 'paradigms' and Polanyi, 'implicit' knowledge) could influence what scientists consider true or not. Early fans of Fort's perspective and work even included the iconoclastic hard-nosed journalist H.L. Mencken.

In the 1960s, American physicist and writer William R. Corliss began his own documentation of scientific anomalies along much more conservative lines than Fort. Corliss claims to be at least partially inspired by Fort and went so far as to check some of Fort's sources. Corliss concluded that Fort left more work to be done with regard to the cataloging of scientific anomalies (Corliss 2002).

> *My second unanticipated discovery made me realize that anomalies were common in all branches of science. This happened in 1953 in the library at the University of Col-*

orado when I was trying to find out what was known about the solar spectrum in the far ultraviolet. (The Physics Department had spectro-grams of the sun taken at high altitudes during flights of captured V-2 German rockets.) Right next to a book I desired was Charles Fort's The Book of the Damned. Naturally, I had to take out that book, too. It turned out to be chock full of anomalies of all sorts, all of which Fort had extracted from major science journals prior to 1930. Fort designated these anomalies as 'damned' because they were generally ignored by mainstream science. (Corliss 2002)

In *Unexplained!* author Jerome Clark explains the difference between Corliss and Fort by saying that Corliss is "more interested in unusual weather, ball lighting, geophysical oddities, extraordinary mirages, and the like–in short, anomalies that, while important in their own right, are far less likely to outrage mainstream scientists than those that delighted Fort, such as UFOs, monstrous creatures, or other sorts of extraordinary events and entities" (Clark 2003, 466-467). For example, Corliss catalogs the claims from both sides of the infinite-dilution debate in chemistry. Homeopathy and medicine have long had an antagonistic relationship. Homeopathy is a form of alternative medicine that treats its patients with dilutions of substances that are believed to cause effects similar to the symptoms presented. These remedies are diluted so much that none of the original substance remains. When the authoritative scientific journal *Nature* published the controversial findings of J. Benveniste that claimed that a solution of antibodies diluted by a factor of 10^{120} triggered a response from 40-60% of the white blood cells tested, the anomalous idea that a remedy diluted beyond Avogadro's constant can have an effect troubled mainstream scientists. For Corliss, the case of 'infinite dilution' should not be closed as "[t]oo many unexplained data survive. We doubt, however, that many scientists will rush to their labs to explore this subject. It would be too risky in the present scientific environment.

Nature has, in effect, relegated 'infinite dilution' research to pseudoscience, whether deserved or not" (Corliss 1988).

While there are many more fascinating anomalies like this in chemistry and physics, it is in the field of psychology that anomalies have recently received much attention. The study of the unusual or paranormal in psychology is called parapsychology, and it has begun to attract serious attention in academic and intelligence circles, although most people aren't aware of the research. ESP (extra-sensory perception) has received the most attention.

One form of ESP that has received vast amounts of study and funding (both from public and private sources) has been 'remote viewing.' Remote viewing is the act of attaining information about some person, place, or thing in particular without engaging one of the five senses. Following the declassification of documents related to the 20 million dollar 'Stargate Project'[1],[2] sponsored by the U.S. Federal Government in the 1990s, ESP surfaced as a subject that was no longer taboo to study. Not all the programs in parapsychology are governmental, however; some are actually thriving in academia. The ubiquitous Rockefeller money funded the Princeton Engineering Anomalies Research (aka PEAR) for many years before they closed their doors. Goldsmiths, University of London, has the Anomalistic Psychology Research Unit[3] and Garret Moddel at University of Colorado, Boulder, offers an 'Edges of Science Course,'[4] among others.

One of the largest private institutions to seriously research parapsychology is SRI International,[5] based in Menlo Park, California.

---

[1] Other such covert research programs sponsored by the CIA went by names like 'Sun Streak' and 'Grill Flame'.

[2] Under the Freedom of Information Act, you can request for STARGATE (remote viewing program) RECORDS that have been released up to the current date. The entire collection totals 89,900 pages in nearly 12,000 documents. For format, please refer to the footnote on page 129.

[3] http://www.gold.ac.uk/apru/.

[4] http://ecee.colorado.edu/~ecen3070/.

[5] http://www.sri.com/about/remoteview.html.

In 1970, the entity was spun off from Stanford University to become an independent non-profit research organization. The U.S. Government funded the psychic research at SRI until 1989. In 1974, two of its research scientists Hal Putoff[6] and Russell Targ published the first full-length peer-reviewed paper on telepathy in *Nature* titled 'Information transfer under conditions of sensory shielding' (Putoff and Targ 1974).

In 1990, government funding for this type of research transitioned to Science Applications International Corporation (SAIC) under the direction of Dr. Edwin May, who had been employed in the SRI program since the mid 1970s and had been Project Director from 1986 until the close of the program.

In 1988, Edwin May and his colleagues analyzed all psi experiments conducted at SRI since 1973. The analysis was based on 154 experiments, consisting of more than 26,000 separate trials, conducted over those sixteen years. Of those, just over a thousand trials were laboratory remote-viewing tests. The statistical results of this analysis indicated odds against chance of 1020 to one (that is, more than a billion billion to one) (Radin 1997, 107). To repeat Mandlebroit's words quoted in the previous chapter,

> *The probability of that happening, ...was less than 105–odds so small they have no meaning. It is a number outside the scale of nature. You could span the powers of ten from the smallest subatomic particle to the breadth of the measurable universe–and still never meet such a number.* (Mandelbrot 2004, 3-4)

In 1995, the US Congress asked two independent scientists to assess whether the $20 million that the government had spent on

---

[6]It is rather interesting that three of the most influential figures within the remote viewing program, Puthoff and remote viewers Ingo Swann and Pat Price, have all achieved the high ranks within L. Ron Hubbard's Scientology system, with Puthoff and Swann achieving the highest rank at the time, Operating Thetan VII.

psychic research had produced anything of value. One of the reviewers was Jessica Utts, a statistics professor at the University of California, Irvine, and an author of textbooks on statistics, who maintained that there had been a statistically significant correlation:

> *It is clear to this author that anomalous cognition is possible and has been demonstrated. This conclusion is not based on belief, but rather on commonly accepted scientific criteria. The phenomenon has been replicated in a number of forms across laboratories and cultures.* (Utts 1995)

The other scientist, Ray Hyman, while skeptical said "I agree with Jessica Utts that the effect sizes reported in the SAIC experiments and in the recent ganzfeld studies probably cannot be dismissed as due to chance" (Hyman 1996). Despite this, the program was shut down in 1995 for failing to find convincing evidence that it had any value to the military or intelligence community.

Are the statistics used to study Psi phenomena creating a Type II error due to bad experiment design (the use of Gaussian distributions to approximate randomness as a benchmark)? Are the social sciences like 'parapsychology' plagued by the same problems as financial modeling? Maybe, our expectations of human abilities aren't wrong; maybe, it is our understanding of randomness and chance that needs reformulating.

The unlikely financial event of Black Monday happened, despite overwhelming odds and so have the positive events in the test for remote viewing. Does this tell us something anomalous about the nature of reality (for example, that ESP exists) or does it tell us that our benchmark of measuring chance (randomness via a Gaussian distribution) is wrong? As outrageous as it may seem, I'd venture to say the latter would generate more opposition than the former since it would require a revision of an enormous amount of our current scientific 'knowledge.' That is to say, I bet that scientists would prefer to include parapsychology under the penumbra of science than revise all the inductive knowledge that has modeled chance using Gaussian distributions.

So if our analogy for chance is misleading, how else are these Psi phenomena being explained away? Famed neuroscientist Michael Persinger has noted correlations between repeated paranormal experiences and geomagnetic phenomena as early as 1985. He researched 25 published cases of profound paranormal activity and their correlations to global geomagnetic activity at the time of their occurrence. Every reported experience happened on days that exhibited geomagnetic activity that was less than the norm for those particular months of the year. The results were "commensurate with the hypothesis that extremely low fields, generated within the earth-ionospheric cavity but disrupted by geomagnetic disturbances, may influence some human behavior" (Persinger 1985).

Persinger went on to create what has come to be called the 'God Helmet,' a helmet equipped with electromagnetic field-emitting solenoids on the sides aimed at the temporal lobes of the wearer.

> *Persinger has tickled the temporal lobes of more than 900 people before me and has concluded, among other things, that different subjects label this ghostly perception with the names that their cultures have trained them to use - Elijah, Jesus, the Virgin Mary, Mohammed, the Sky Spirit. Some subjects have emerged with Freudian interpretations - describing the presence as one's grandfather, for instance - while others, agnostics with more than a passing faith in UFOs, tell something that sounds more like a standard alien-abduction story.* (Hit 1999)

So it seems that there may be some sort of connection between the paranormal and EMF or geomagnetic field (GMF) activity. However, the questions persist: do electromagnetic fields create hallucinations or augment our perceptions, yielding way to a deeper level of awareness? Are psychic phenomena happening more often than what chance would predict or is our idea of chance wrong? Again, a position other than agnosticism at this point seems overreaching, even arrogant. Noted author and proponent of scientific skepticism Sam Harris writes:

> *While there have been many frauds in the history of parapsychology, I believe that this field of study has been unfairly stigmatized. If some experimental psychologists want to spend their days studying telepathy, or the effects of prayer, I will be interested to know what they find out. And if it is true that toddlers occasionally start speaking in ancient languages (as Ian Stevenson alleges), I would like to know about it. However, I have not spent any time attempting to authenticate the data put forward in books like Dean Radin's The Conscious Universe or Ian Stevenson's 20 Cases Suggestive of Reincarnation. The fact that I have not spent any time on this should suggest how worthy of my time I think such a project would be. Still, I found these books interesting, and I cannot categorically dismiss their contents in the way that I can dismiss the claims of religious dogmatists.* (Harris 2009)

Radin is one of a cadre of statistically trained parapsychologists shaping a paradigm that is attempting to integrate psychic phenomena into mainstream science. Many of these parapsychologists leverage the ideas of quantum mechanics to validate their new paradigm, in particular the concept of a Zero-Point Field. In her book *The Field*, bestselling author Lynne McTaggart talks about Zero-Point energy. In it, she makes the argument that since everything in the universe is connected and we too are part of this vast dynamic web of energy exchange, supernatural phenomena make sense 'scientifically.' Radin is the standard bearer here.

> *Scientist Dean Radin says it very succinctly: 'The fact that quantum objects can become entangled means that*

the common sense assumption that ordinary objects are entirely and absolutely separate is incorrect.'

Quantum theory implies that the universe is a single integrated system containing innumerable subsystems. Everything in it is 'entangled' with everything else. But what's so 'spooky' about that? It is, after all, what the word 'universe' means. It's only 'spooky' if the idea of being a part of a larger entity is disturbing to you.

We are trained from birth to see things as disconnected. Our language does it. Learning is as much learning NOT to see as it is learning to see. What a child sees initially is an undifferentiated whole. By careful training it learns to carve pieces out of that reality and look at them as separate material objects. But why should that be considered more 'real'? It's just one way of seeing. People with greater ability to communicate telepathically aren't 'gifted'—they simply haven't been as thoroughly indoctrinated. Instead of asking why some people (perhaps everyone at birth) can communicate telepathically, we should be studying the mechanism that enables us to shut out that information most of the time. (Slater 2009)

Parapsychologists like Radin have been accused of abusing the anomalies, enigmas, and confusing nature of quantum mechanics into some sort of explanation of psi phenomena (for example, see James Alcock's *Parapsychology's Past Eight Years: A Lack-of-Progress Report* (1984)). Again, extraordinary claims require extraordinary proof.

More generally, we have learned that our colleagues' tolerance for any kind of theorizing about psi is strongly determined by the degree to which they have been convinced by the data that psi has been demonstrated. We have further learned that their diverse reactions to the data

> *themselves are strongly determined by their a priori beliefs about and attitudes toward a number of quite general issues, some scientific, some not. In fact, several statisticians believe that the traditional hypothesis testing methods used in the behavioral sciences should be abandoned in favor of Bayesian analyses, which take into account a person's a priori beliefs about the phenomenon under investigation* (e.g., Bayarri & Berger, 1991; Dawson, 1991).
>
> *In the final analysis, however, we suspect that both one's Bayesian a prioris and one's reactions to the data are ultimately determined by whether one was more severely punished in childhood for Type I or Type II errors.* (Dem and Honorton, 1994)

The anomalist opposes dogmatic centralization in thought, power, resources, and influence–centralization in thought in the dogmas of religion, scientism, psychiatry, and so on; centralization of power in central or federal governments; centralization of wealth in central banks; centralization in influence through wealth and power; and centralization in education through government control. The anomalist is skeptical of skepticism and as such seeks evidence that challenges the status quo. This isn't nihilistic; it is creative destruction like the demolition of an old building in favor of a newer, more functional and accommodating one. This isn't relativism since the anomalist doesn't hold that all belief systems are equal. Ultimately, the anomalist stands opposed to the problems presented by the centralization of symbology and isn't afraid to present evidence that may subvert the dominant paradigm, whether it be a scientific, political, or cultural one. As David Hume points out, a "wise man, therefore, proportions his belief to the evidence."

> *The system of concepts is a creation of a man together with the rules of syntax, which constitutes the structure of all conceptual systems. ...All concepts, even those which*

are closest to experience, are from the point of view of logic freely chosen conventions, just as is the case with the concepts of causality... (Albert Einstein)

What does all this mean? It means that anomalies can be indicators of truth and reality; they are not to be shirked but studied and integrated. Given mankind's penchant for trope, constant vigilance is needed to protect our right to discuss, investigate, and know our world.

RECOMMENDED RESOURCES

'The Structure of Scientific Revolutions' by Thomas S. Kuhn – This is a classic in the philosophy of science and offers a powerful framework for the progress of knowledge in the same way that evolution did for biology.

'The Mind's I' composed and arranged by Douglas R. Hofstadter and Daniel C. Dennett – A wonderfully readable book put together by two powerful and insightful thinkers.

'War Against The Weak' by Edwin Black – A very academic and thorough history of the eugenics movement. Highly recommended.

'The Misbehavior of Markets' by Benoit Mandelbrot – This needs to be required reading for every graduate student in finance.

'Science Frontiers' by William Corliss – Corliss' compendium of anomalies is as fun to read as it is comprehensive.

'Husband-Coached Childbirth' by Robert A. Bradley, M.D. – For those concerned about the high infant mortality statistics in the United States, the Bradley Method of Husband Coached Childbirth is really an eye-opener.

'Psychiatry: The Science of Lies' and 'Antipsychiatry: Quackery Squared' by Thomas Szasz – Thomas Szasz has been central to the debate about mental illness for half a century. These books should be required reading.

'The Blue Sense' by Arthur Lyons and Marcello Truzzi, Ph. D. – While this is an older book, it is a model for how investigative journalism with regards to psychic phenomena should be done.

'Science and the Founding Fathers' by I. Bernard Cohen – Anyone interested in epistemology and the American Revolution will love this book. Wonderful nuggets of wisdom sprinkled throughout.

'Power and Market' by Murray N. Rothbard – Rothbard is an interesting thinker with some very strong arguments and clever ideas. In particular, he touches upon how wealth can be decentralized by simply revoking limited liability for corporations. Policy makers looking for creative solutions would be smart to be inspired by Rothbard.

'The C.I.A. Doctors' by Colin Ross – A wonderful book that uses Freedom of Information Act documents as source material which details human rights abuses by psychiatrists who intentionally created dissociative identity disorder (that is, multiple personalities) in patients.

'The Field' by Lynne McTaggart – An impassioned (although some may prefer the word 'biased') account of the attempt to reconcile multiple anomalies in physics, psychology, and chemistry via the quantum concept of the Zero-Point Field. Highly recommended.

'JFK and The Unspeakable' by James W. Douglass – This is one of the most up-to-date and thorough accounts of the assassination of President Kennedy. Picked up a copy after former senior CIA analyst Ray McGovern praised it's veracity and scholarship.

'The Conscious Universe' by Dean Radin, Ph.D. – A rigorous yet accessible account of the current Psi research and its implications. This book may blow your mind.

'The Best Way to Rob a Bank is to Own One' by William K. Black – A wonderful analysis of the causes of our current financial crisis of confidence from a former Senior regulator during 1980s S&L debacle.

'The Family' by Jeff Sharlet – A shocking account by respected journalist Sharlet of the 'secret' organization influencing the world's leaders and their beliefs.

'The Prosecution of George W. Bush for Murder' by Vincent Bugliosi – A book that deserves wider readership, that holds the executive branch accountable for its actions. Bugliosi is famous for successfully prosecuting Charles Manson.

'The Paleo Diet' by Loren Cordain – Truly original and convincing research drives Professor Loren Cordain's diet program. This is required reading.

'The UltraMind Solution' by Mark Hyman, M.D. – A great book that extols the virtues of proper nutrition, i.e., preventive medicine, in lieu of prescribing medications to deal with symptoms of illness.

'Body of Secrets' by James Bamford – This is an eye-opening account of perhaps the most powerful Black Chamber in the history of humankind.

'The Franklin Cover-Up' by Senator John Decamp – A shocking book by a U.S. Senator that itself seems to be the victim of the very cover-up it is trying to expose.

'Tricks of the Mind' by Derren Brown – Derren Brown is doing excellent work jolting people into questioning the dominant paradigms while fostering a healthy love of critical thinking.

'Business of Being Born' produced by Ricki Lake – This documentary needs to be seen by every pregnant woman in the United States during her first trimester.

'Generation Rx' produced by Kevin P. Miller – This well researched documentary illustrates the abuses and horrific consequences of the mass drugging of our youth.

'Endgame' produced by Alex Jones – Alex Jones is quite controversial but to ignore him and his bullhorn is foolish. His radio program and prolific DVD production are to be lauded for his respect for the Constitution of the United States.

'The PEAR Proposition: Scientific Study of Consciousness-Related Physical Phenomena' – This wonderful DVD gives Dr. Robert Jahn and his associates ample time to explain their methodology and results in both Remote Viewing and Human-Machine interaction and the role of intention in randomly generated events. Highly recommended.

The Citizen's Commission on Human Rights – A powerful psychiatry watchdog, that while being co-founded by Scientology, is wholly secular and with many members from many religious and non-religious backgrounds (including Thomas Szasz). Highly recommended http://www.CCHRint.org

Electronic Freedom Foundation – A great group of individuals working for liberty in cyberspace. Check this particular link to learn how to communicate with your Congressperson http://w2.eff.org/congress

Recommended Resources

Mind Hacks blog – This is an excellent blog for those interested in the mind and the latest research in cognitive neuroscience http://www.mindhacks.com/

Furious Seasons blog – Another stellar blog that documents and exposes the abuses of big Pharma http://www.furiousseasons.com/

Skeptical Investigations – A wonderful resource for those more interested in open minded skepticism than the close minded variety http://skepticalinvestigations.org/

Society for Scientific Exploration – This site is home to many avant-garde scientists and engineers willing to discuss unusual and anomalous phenomenon. Highly recommended. http://www.scientificexploration.org/

BIBLIOGRAPHY

Journal American Medical Association 284, no. 4 (July 2000): 483.

Taber's Medical Dictionary. F A Davis Co, February 2009.

Adams, John. "Letter to Thomas Jefferson." September 3, 1816.

Adams, John, and Thomas Jefferson. "Draft of the Declaration of Independence." *Wikisource.* http://en.wikisource.org/wiki/Draft_of_the_Declaration_of_Independence (accessed October 09 2009).

Alcock, James. "Parapsychology's Past Eight Years: A Lack-of-Progress Report." *Skeptical Inquirer* 8, no. 312 (1984).

Allen, Thomas B. *Declassified: 50 Top-Secret Documents That Changed History.* National Geographic, 2008.

Ansell-Pearson, Keith. *An Introduction to Nietzsche as Political Thinker.* Cambridge: Cambridge University Press, 1994.

Arrow, Kenneth. "A Difficulty in the Concept of Social Welfare." *Journal of Political Economy* 58, no. 4 (August 1950): 328-346.

Associated Press, CNN. *Atheist soldier claims harassment.* April 26, 2008. http://www.militaryreligiousfreedom.org/press-releases/

ap_cnn_atheist.html.

Bachelier, Louis. *The Theory of Speculation*. 1900.

Bagdikian, Ben. *The Media Monopoly*. Boston: Beacon Press, 1983.

Ball, Ray. "The Theory of Stock Market Efficiency: Accomplishments and Limitations." In *The New Corporate Finance; Where Theory Meets Practice,* edited by Jr. D. H. Chew, 35-48. Boston, MA, 1994.

Banerjee, Neela. *Soldier Sues Army, Saying His Atheism Led to Threats*. April 26, 2008. http://www.nytimes.com/2008/04/26/us/26atheist.html?_r=1 (accessed September 19, 2009).

Baudrillard, Jean. *The Gulf War Did Not Take Place*. Bloomington: Indiana University Press, 1995.

Bausell, R. Barker. *Snake Oil Science*. New York: Oxford University Press, 2007.

Bealle, Morris A. *The New Drug Story.* Washington, D.C.: Columbia Publishing Co., 1958.

Bem, Daryl J. and Honorton, Charles (1994). "Does Psi Exist?". *Psychological Bulletin*, Vol. 115, No. 1, 4-18. http://www.dina.kvl.dk/~abraham/psy1.html.

Biderman, Albert D., and Herbert Zimmer. "Introduction." In *The Manipulation of Human Behavior*, edited by Albert D. Biderman and Herbert Zimmer, 6. New York: John Wiley & Sons, 1961.

Black, Edwin. *The War Against the Weak*. New York, London: Four Walls Eight Windows, 2003.

Bibliography

Blass, Dr. Thomas. "The Milgram paradigm after 35 years: Some things we now know about obedience to authority." *Journal of Applied Social Psychology* 29, no. 5 (1999): 955-978.

Bly, Nellie. "Tormenting the Insane." *The New York Times*, 1879.

Boorstin, Daniel J. *Hidden History.* New York: First Vintage Book, 1961.

Borge, Dan. *The Book of Risk.* New York: John Wiley & Sons, Inc., 2001.

Braid, James. *Neurypnology.* London: J Churchill, 1853.

Brenman, M. "Experiments in the hypnotic production of anti-social and self-injurious behavior." *Psychiatry* 5 (1942): 49-61.

Brinker, Menahem. "Nietzsche and the Jews." In *Nietzsche, Godfather of Fascism?*, edited by Jacob Golomb and Robert S Wistrich, 107-125. Princeton, New Jersey: Princeton University Press, 2002.

Browne, Harry. *How I Found Freedom in an Unfree World.* LiamWorks, 1998.

Bulwer-Lytton, Edward. *Richelieu; Or the Conspiracy: A Play in Five Acts.* Conduit St., London: Saunders and Otley, 1839.

Burke, Kenneth. *A Rhetoric of Motives.* Berkeley, California: University of California Press, 1962.

Butler, Smedley. *Common Sense*, 1935.

—. *War is a Racket.* 1935.

Carpenter, Ted Galen. *The Captive Press: Foreign Policy Crises and the First Amendment.* Washington, D.C.: The Cato Institute, 1995.

Carrel, Alex. *L'Homme, cet Inconnu (Man, the Unknown).* New York and London: Harper and Brothers, 1935.

Center for Responsive Politics. *Top All-Time Donors 1989-2010 Summary.* August 23, 2009. http://www.opensecrets.org/orgs/list.php (accessed October 08, 2009).

Chesterton, G. K. *The Ball and the Cross.* Kessinger Publishing, 2004.

Chisholm, G. Brock. "Mental Health and World Citizenship." Broadway, New York: Distributed by the National Association For Mental Health, Inc., 1790. 7-8.

Chomsky, Noam. *Language and Mind.* New York: Cambridge UP, 2006.

Clark, Jerome. *Unexplained!* Detroit: Visible Ink Press, 2003.

Cline, Andrew R. *Tropes and Schemes.* 2006. http://rhetorica.net/tropes.htm (accessed September 12, 2009).

Cohen, I. Bernard. *Science and the Founding Fathers.* New York: W.W. Norton and Company, Inc., 1997.

Condorcet, Marquis de. *Essay on the Application of Analysis to the Probability of Majority Decisions.* 1785.

Corliss, William R. "A Search for Anomalies." *Journal of Scientific Exploration* 16, no. 3 (2002): 439-453.

Bibliography

Corliss, William R. "Update on the "infinite dilution" experiments." *Science Frontiers (Online)*, no. 60 (November-December 1988).

Crystal, David. *How Language Works*. New York: Avery (Penguin), 2005.

Dawar, Anil. "Prozac: Found in Tapwater." *Daily Mail*, August 9, 2004.

Dawkins, Richard. *The Selfish Gene*. Oxford, New York: Oxford University Press, 1976.

Derman, Emanuel. "What Quants Don't Learn at College." *Risk Magazine*, July 2003.

Descartes, René. *Meditations on First Philosophy*. Edited by John Cottingham. Cambridge University Press, 1996.

Doren, Charles Van. *The History of Knowledge*. New York: Ballantine Books, 1991.

Dreman, David N., and Michael A. Berry. "Overreaction, Underreaction, and the Low-P/E Effect." *Financial Analysts Journal* 51, no. 4 (1992): 21-30.

Edwards, Paul, ed. *The Encyclopedia of Philosophy*. Vol. I. New York: Macmillan Publishing Co., Inc. and The Free Press, 1967.

Eggen, Dan. *9/11 Panel Suspected Deception by Pentagon*. August 2, 2006. http://www.washingtonpost.com/wp-dyn/content/article/2006/08/01/AR2006080101300.html (accessed September 30, 2009).

Empiricus, Sextus. *Outlines of Pyrrhonism*. Translated by R.G. Bury. London: W. Heinemann, 1933.

Erickson, Milton. "An Experimental Investigation of the Possible Anti-Social Use of Hypnosis." *PSYCHIATRY* 2 (1939).

Estabrooks, Dr. George. *Hypnotism*. New York: E.P. Dutton, 1944.

Estabrooks, George. "George Estabrooks Writing to Milton Erickson." In *The Letters of Milton H. Erickson*. Phoenix: Zeig, Tucker & Theisen, Inc., 2000.

Evans, Dylan. *Placebo*. New York: Oxford University Press, 2004.

Fama, Eugene F., and Kenneth R. French. "Dividend Yields and Expected Stock Returns." *Journal of Financial Economics* 3-25 (October 1988).

Fama, Eugene. "The Behavior of Stock Market Prices." *Journal of Business* 38, no. 1 (January 1965): 34-105.

Federal Reserve Bank of Chicago. *Modern Money Mechanics: A Workbook on Deposits, Currency and Bank Reserves*. Federal Reserve Bank of Chicago, 1971.

Federal Reserve Bank of St. Louis. "A Foregone Conclusion, Chapter 2: Banking Reform 1907 - 1913." *Federal Reserve Bank of St. Louis*. http://www.stlouisfed.org/foregone/chapter_two.cfm#thirteen (accessed October 10, 2009).

Festinger, Leon. *When Prophecy Fails: A Social and Psychological Study of a Modern Group that Predicted the Destruction of the World*. University of Minnesota Press, 1956.

Bibliography

Fingarette, Herbert. *Self-Deception.* Berkeley and Los Angeles: University of California Press, 2000.

Fosdick, Raymond Blaine. *The Story of the Rockefeller Foundation.* New Brunswick: Transaction Publishers, 1989.

French, Kenneth R., and Richard Roll. "Stock return variances: The arrival of information and the reaction of traders." *Journal of Financial Economics* (Elsevier) 17, no. 1 (September 1986): 5-26.

Galilei, Galileo. *Il saggiatore (The Assayer).* 1864.

Gatto, John Taylor. "Some Lessons from the Underground History of American Education." *The Odysseus Group.* 2000. http://www.johntaylorgatto.com/chapters/index.htm (accessed October 13, 2009).

Gilbert, G.M. *Nuremberg Diary.* York: Da Capo Press, Inc., 1947.

Gladwell, Malcolm. *Outliers.* New York: Little, Brown, and Company, 2008.

Glattfelder, James, and Stefano Battiston. "Backbone of complex networks of corporations: The flow of control." *Physical Review E* 80, no. 3 (2009).

Goldwater, Barry. *With No Apologies.* Morrow, 1979.

Gross, Alan G. *The Rhetoric of Science.* Cambridge, Massachusetts and London: Harvard University Press, 1990.

Grossman, Dave. *On Killing.* New York, London: Back Bay Books; Little, Brown and Company, 1995.

Hacohen, Malachi Haim. *Karl Popper–The Formative Years, 1902-1945*. Cambridge, Massachusetts: Cambridge University Press, 2002.

Harris, Sam. "Response to Controversy." *Sam Harris*. August 11, 2009. http://www.samharris.org/site/full_text/response-to-controversy2/ (accessed October 10, 2009).

—. *The End of Faith*. New York: W.W. Norton & Company, 2005.

Hawking, Stephen. *A Brief History of Time*. Bantam Books, 1988.

Hayek, F. A. "The Use of Knowledge in Society." *The American Economic Review* XXXV, no. 4 (September 1948).

Hayek, F. A. "Why I Am Not A Conservative." In *The Constitution of Liberty*. Chicago: The University of Chicago, 1960.

Hit, Jack. "This is Your Brain on God." *Wired (Online)*, November 1999.

Hitler, Adolf. *The Speeches of Adolf Hitler Volume I, April 1922-1939*. Edited by Norman H. Baynes. Oxford, UK: Oxford University Press, 1942.

Hofstadter, Douglas R. "Analogy as the Core of Cognition." In *The Analogical Mind: Perspectives from Cognitive Science*, edited by Dedre Gentner, Keith J Holyoak and Boicho N. Kokinov, 499-538. Cambridge MA: The MIT Press/Bradford Book, 2001.

Hofstadter, Douglas. *The Mind's I*. New York: Bantam, 1982.

Hume, David. *An Enquiry Concerning Human Understanding*. The Secular Web: http://www.infidels.org/library/historical/david_

hume/human_understanding.html, 1748.

Hyman. "Evaluation of a program on anomalous mental phenomena." *Journal of Scientific Exploration*, 1996 : 39-40.

International Military Tribunal. "Trial of The Major War Criminals before the International Military Tribunal, Nuremberg, Volume 9." 1945.

Jakobson, Roman. *What is Poetry?* 1933.

Jaynes, Julian. *The Origins of Consciousness in the Break Down of the Bicameral Mind.* Boston and New York: Mariner Books (Houghton Mifflin Company), 2000.

Jefferson, Thomas. "Letter to Albert Gallatin, Secretary of the Treasury." 1802.

—. "Letter to Dr. Thomas Cooper." February 10, 1814.

—. *Notes on Virginia.* 1782.

Jones, Jeffrey M. "Americans Have Net-Positive View of U.S. Catholics." *Gallup.* April 15, 2008. http://www.gallup.com/poll/106516/americans-netpositive-view-us-catholics.aspx (accessed October 09, 2009).

Joseph, Sister Miriam. *The Trivium: The Liberal Arts of Logic, Grammar, and Rhetoric.* Philidelphia: Paul Dry Books, Inc., 2002.

Kahn, David. *The Codebreakers: The Comprehensive History of Secret Communication from Ancient Times to the Internet.* New York: Scribner, 1967.

Kant, Immanuel. *Critique of Pure Reason.*

Kelves, Daniel J. *In the Name of Eugenics.* Cambridge & London: Harvard University Press, 1995.

Kirsch, Irving, Brett J. Deacon, Tania B. Heudo-Medina, Alan Scoboria, Thomas J. Moore, and Blair T. Johnson. "Intial Severity and Antidepressant Benefits: A Meta-Analysis of Data Submitted to the Food and Drug Administration." *PLoS Medicine* (Public Library of Science), February 2008: http://www.plosmedicine.org/article/info:doi/10.1371/journal.pmed.0050045.

Korzybski, Alfred. *Science and Sanity: An Introduction to Non-Aristotelian Systems and General Semantics.* Institute of General Semantics, 1958.

Kuhn, Thomas S. *The Essential Tension: Selected Studies in Scientific Tradition and Change.* University of Chicago Press, 1977.

—. *The Structure of Scientific Revolutions.* Chicago: University of Chicago Press, 1970.

Lakoff, George, and Mark Johnson. *Metaphors We Live By.* Chicago: University of Chicago Press, 1980.

Lang, Berel. "Misinterpretation as the Author's Responsibility (Nietzsche's fascism, for instance)." In *Nietzsche, Godfather of Fascism?*, edited by Jacob Golomb and Robert S. Wistrich, 47-65. Princeton, New Jersey: Princeton University Press, 2002.

Lanham, Richard A. *The Economics of Attention.* Chicago and London: University of Chicago Press, 2006.

Bibliography

Leonhardt, David. *Medical Malpractice System Breeds More Waste.* September 22, 2009. http://www.nytimes.com/2009/09/23/business/economy/23leonhardt.html? (accessed October 08, 2009).

Lewis, C.S. *God in the Dock.* 1970.

Libet, Benjamin. *Mind time: The temporal factor in consciousness.* Cambridge, MA: Harvard University Press, 2004.

Lippman, Walter. *Public Opinion.* University of Virginia American Studies Program, http://xroads.virginia.edu/~Hyper2/CDFinal/Lippman/cover.html, 2003.

Locke, John. *An Essay Concerning Human Understanding.* Edited by Roger Woolhouse. New York: Penguin Books, 1997.

Mach, Ernst. "The Economical Nature of Physical Inquiry." In *Philosophy of Science: The Historical Background*, by J. Kockelmans. New York: The Free Press, 1968.

Madison, James. "Memorial and Remonstrance against Religious Assessments." 1785.

—. *The Federalist Papers, No. 10.* New York: NAL Penguin, 1961.

Malkiel, B.G. *A Random Walk Down Wall Street.* New York, 1996.

Mandelbrot, Benoit. *The (Mis)behavior of Markets.* New York: Basic Books, 2004.

Mandeville, Bernard. "The Grumbling Hive." In *The Fable of the Bees.* London: Penguin Group, 1989.

Markowitz, Harry. "Portfolio Selection." *Journal of Finance* 7, no. 1 (1952): 77-91.

Mauthner, Fritz. *Dictionary of Philosophy: New Contributions to a Critique of Language*. Vol. 1. Leipzig: Felix Meiner, 1910-11.

McClosky, Deirdre. *The Rhetoric of Economics*. University of Wisconson Press, 1998.

Mencken, H.L. *A Mencken Chrestomathy.* New York: Alfred A. Knopf, 1956.

Milgram, Stanley. *Obedience to Authority: An Experimental View.* New York: Harper and Row Publishers Inc., 1974.

Miller, Mark Crispin. "Our Rigged Elections." *The Baltimore Chronicle*, October 4, 2006.

Miller, Merle. *Plain Speaking: An Oral Biography of Harry S. Truman.* 1973.

Mindrum, Michael R. "Time for another revolution? The Flexner Report in historic context, reflections on our profession." *Coronary Artery Disease* 17, no. 5 (August 2006): 477-481.

Morris, Roger. *Partners in Power.* 1996.

Morville, Peter. *Ambient Findability: What We Find Changes Who We Become.* O'Reilly Media, Inc., 2005.

Müller-Lauter, Wolfgang. "Experiences with Nietzsche." In *Nietzsche, Godfather of Fascism?*, edited by Jacob Golomb and Robert S Wistrich, 66-89. Princeton, New Jersey: Princeton University Press, 2002.

Bibliography

Neisser, Ulric, and Nicole Harsch. "Phantom Flashbulbs." In *Memory observed: Remembering in Natural Contexts*. New York: Worth Publishers, 2000.

Nietzsche, Friedrich. "From Ecce Hommo: How One Becomes What One Is." In *Philosophical Writings*, edited by Reinhold Grimm and Caroline Molina y Vedia. New York: The Continuum Publishing Company, 1997.

———. *Human, All Too Human*. Translated by R.J. Hollingdale. Cambridge: Cambridge University Press, 1986.

———. "On Truth and Lies in a Nonmoral Sense." 1873.

———. *Thus Spoke Zarathustra: A Book for Everyone and No one*. Penguin Books, 1961.

Oath of Office. http://www.senate.gov/artandhistory/history/common/briefing/Oath_Office.htm.

Olfson, Mark, and Steven Marcus. "National Patterns in Antidepressant Medication Treatment." *Archives of General Psychiatry* 66, no. 8 (2009): 848-856.

Oppenheimer, Mark. *The Times*, September 19, 2008.

Orwell, George. *1984*. New York: Signet Classics, 1950.

Paine, Thomas. *The Age of Reason*. The Secular Web: http://www.infidels.org/library/historical/thomas_paine/age_of_reason/, 1975.

Pareto, Vifredo. *La transformazioni della democrazia*. Edited by Mario Missiroli. Capelli Editore, 1964.

Persinger, Michael A. "Geophysical Variables and Behavior:XXX. Intense Paranormal Experiences Occur during Days of Quiet, Global, Geomagnetic Activity." *Perceptual and Motor Skills* 61, no. 320 (1985).

Pert, Candace B. "Letter to the Editor." *TIME Magazine.* October 29, 1997.

Pew Research Center. *Survey Reports.* September 13, 2009. http://people-press.org/report/543/.

Pilger, John. http://www.youtube.com/watch?v=C62KAmMzu0E, 2009.

Pohl, Rudiger. *Cognitive Illusions: A Handbook on Fallacies and Biases in Thinking, Judgment, and Memory.* Hove and New York: Psychology Press, 2004.

Polyani, Michael. *Personal Knowledge.* Routledge & Kegan Paul Ltd, 1962.

Postel, Gert. *Gert Postel: German Television Interview.* October 26, 2001. http://www.gert-postel.de/english.htm (accessed October 14, 2009).

Putoff, Hal, and Russell Targ. "Information transfer under conditions of sesory shielding." *Nature* 252, no. 5476 (October 1974): 602-607.

Quigley, Carroll. *Tragedy and Hope: A History of the World in Our Time.* San Pedro: G. S. G. & Associates, Inc., 1975.

Radin, Dean. *The Conscious Universe.* New York: HarperCollins, 1997.

Rees, John Rawlings. "Strategic Planning for Mental Health." *Mental Health* 1, no. 4 (October 1940).

Reiter, Dr. Paul J. *Antisocial or Criminal Acts of Hypnosis: A Case Study.* Munksgaard, 1958.

Richman, Sheldon. *Separating School and State.* Fairfax, VA: Future of Freedom Foundation, 1995.

Rocco, Alfredo. *The Political Doctrine of Fascism.* Translated by Dino Bigongiari. Carnegie Endowment for International Peace, Division of Intercourse and Education, 1926.

Rockfeller, David. *Memoirs.* New York: Random House, Inc., 2002.

Rosenhan, D.L. "On Being Sane in Insane Places." *Science* (American Association for the Advancement of Science) 179, no. 4070 (January 1973): 250–258.

Rosenhan, D.L. "On Being Sane in Insane Places." *Science*, 1973: 250-258.

Rothbard, Murray N. "The Origins of the Federal Reserve." *The Quarterly Journal of Austrian Economics* 2, no. 3 (Fall 1999): 3-51.

Rothbard, Murray N., and Hans-Herman Hoppe. *The Ethics of Liberty.* NYU Press, 2003.

Rowland, Loyd. "Will Hypnotized Persons Try To Harm Themselves or Others?" *Journal of Abnormal and Social Psychology* 34 (1939): 114-117.

Russell, Bertrand. *History of Western Philosophy.* New York: Simon and Schuster, 1945.

Russell, Bertrand. "Is There a God?" In *The Collected Papers of Bertrand Russell: Volume 10*, edited by John Slater and Peter Köllner, 542-548. Routledge, 1996.

—. *The Impact of Science on Society.* London and New York: Routledge, 1952.

—. *Why I Am Not A Christian.* 1957.

Sargant, William Walters. *The Mind Possessed: A Physiology of Possession, Mysticism, and Faith Healing.* Lippincott, 1974.

Schivelbusch, Wolfgang. *Three New Deals.* New York: Picador, 2006.

Segal, Nancy. *Entwined Lives: Twins and What They Tell Us About Human Behavior.* New York: Plume Books, 2000.

Sharlet, Jeff. "Jesus plus nothing: Undercover about America's secret theocrats." *Harper's magazine.* March 2003. http://www.harpers.org/archive/2003/03/0079525 (accessed October 09, 2009).

Shermer, Michael. "Patternicity." *Scientific American*, December 2008.

Shermer, Michael. "The Shamans of Scientism." *Scientific American*, June 2002.

Shiller. "Human Behavior and the Efficiency of the Financial System." 1997.

Bibliography

Shirer, William L. *The Rise and Fall of the Third Reich: A History of Nazi Germany.* New York: Touchstone, 1981.

Slater, Philip. "Why What Frightens 'Skeptics' Frightened Einstein." *The Huffington Post.* April 8, 2009. http://www.huffingtonpost.com/philip-slater/why-what-frightens-skepti_b_184778.html (accessed Ocober 10, 2009).

Smith, Page. *Democracy on Trial.* New York: Simon & Schuster, 1995.

Steigmann-Gall, Richard. *The Holy Reich: Nazi Conceptions of Christianity.* Cambridge: Cambridge University Press, 2003.

"Strategic Planning for Mental Health." June 1940.

Szasz, Thomas. *Myth of Psychotherapy.* Syracuse University Press, 1988.

—. *Pharmacracy: Medicine and Politics in America.* Westport, Connecticut: Praeger Publishers, 2001.

—. *Psychiatry: The Science of Lies.* New York: Syracuse University Press, 2008.

Taleb, Nassim Nicholas. *The Black Swan.* New York: Random House, 2007.

The American Psychiatric Association. *The Diagnostic and Statistical Manual of Mental Disorders.* Fourth Edition, Text Revision. Washington, D.C.: The American Psychiatric Association, 2000.

"Treasury Bond Auctions Show Insatiable Debt Demand."

Bloomberg.com. September 11, 2009.

"Treaty of Tripoli, Article 11." 1796.

Truzzi, Marcello. "Anomalistics: The Perspectives of Anomalistics." *Skeptical Investigations.* 1998. http://skepticalinvestigations.org/anomalistics/perspective.htm (accessed October 10, 2009).

—. "On Pseudo-Skepticism." *Zetetic Scholar*, 1987: 12-13.

Twight, Charlotte. *America's Emerging Fascist Economy.* New Rochelle, New York: Arlington House, 1975.

United States Congress, House of Representatives, House Committee on Un-American Activities. *Investigation of Nazi Propaganda Activities and Investigation of Certain Other Propaganda Activities: Public Hearings before the Special Committee on Un-American Activities.* 73rd Congress, 2nd Session, Hearings No. D.C. 6II, Washington, D.C.: Governement Printing Office, 1935.

US House of Representatives, Subcommittee of the Committee on Banking and Currency. *Hearings before the Subcommittee of the Committee on Banking and Currency, House of Representatives, 72nd Congress, 1st Session, on H.R. 10517, for Increasing and Stabilizing the Price Level of Commodities and for Other Purposes.* Washington: Government Printing Office, 1932.

Utts, Jessica. "An Assessment of the Evidence for Psychic Functioning." *UCDavis: University of California–Department of Statistics.* 1995. http://anson.ucdavis.edu/~utts/air2.html#copyright (accessed October 10, 2009).

Vickers, John. *The Problem of Induction.* September 08, 2009. http://plato.stanford.edu/entries/induction-problem/.

Wells, W.R. "Experiments in the hypnotic production of crime." *Journal of Psychology* 11 (1941): 63-102.

Wells, Wesley. "Wesley Wells Writing to Milton Erickson." In *The Letters of Milton H. Erickson,* 228. Phoenix: Zeig, Tucker & Theisen, Inc., 2000.

Wilde, Oscar. *The Soul of Man Under Socialism.* Vol. IV, in *The Complete Works of Oscar Wilde: Historical Criticism,* edited by Josephine M. Guy. USA: Oxford University Press, 2007 (reprint edition).

Wilson, Woodrow. *The New Freedom, A Call For the Emancipation of the Generous Energies of a People.* New York and Garden City: Doubleday, Page & Company, 1913.

Wittgenstein, Ludwig. *Philosophical Investigations.* Translated by G.E.M. Anscombe, Hacker, P.M.S. and Schulte Joachim. Wiley-Blackwell, 2009.

—. *Tractatus Logico-Philosophicus.* Translated by D.F. Pears and B.F. Mc Guinness. London: Routledge and Kegan Paul, 1961.

Wroughton, Lesley. "SNAP ANALYSIS: New world economic order takes shape at G20." *Reuters.* September 25, 2009. (http://www.reuters.com/article/ousivMolt/idUSTRE58O1FB20090925?virtualBrandChannel=11604) (accessed October 13, 2009).

ABOUT JAKE SHANNON

"There he goes. One of God's own prototypes. Some kind of high powered mutant never even considered for mass production. Too weird to live, and too rare to die." – Hunter S. Thompson

Jake Shannon is a family man, acclaimed author, professional hypnotist, financial engineer, physical culturalist, Human Rights investigator, inventor, entrepreneur, and a teenage cancer survivor. In 1989, upon his oncologist's recommendation, Jake started using hypnosis and visualization to overcome the pain of radiation and chemotherapy. By 1992, Jake was using his proprietary combination of hypnosis and critical thinking to achieve his personal, professional and academic goals. Jake's academic accomplishments span a broad domain—he has a Bachelor of Arts degree in English and a Master of Science in Financial Engineering. In 2007, Jake turned fulltime to helping others in diverse fields, such as finance, persuasion, fitness, and entrepreneurship. He founded scientificwrestling.com, invented the full-body fitness tool the Macebell, and co-founded the innovative reverse mortgage consultancy Reverse Mortgage Insight. He also undertakes lectures and is a featured guest on talk radio programs nationwide. For more information, please visit http://www.JakeShannon.com.

Anomaly: Revolutionary Knowledge in Everyday Life

Made in the USA
Charleston, SC
21 September 2010